Y0-ACF-773

Publication No. 21
of the Mathematics Research Center
United States Army
The University of Wisconsin

ERROR CORRECTING CODES

A message with content and clarity
Has gotten to be quite a rarity.
    To combat the terror
    Of serious error,
Use bits of appropriate parity.

Solomon W. Golomb

ERROR CORRECTING CODES

Proceedings of a Symposium
Conducted by the Mathematics Research Center,
United States Army at the
University of Wisconsin, Madison
May 6–8, 1968

Edited by

Henry B. Mann

JOHN WILEY & SONS, INC.
New York • London • Sydney • Toronto

Library of Congress Catalog Card Number: 69–15487
SBN 471 56715 9
Printed in the United States of America

FOREWORD

This volume is the proceedings of a symposium on error correcting codes organized by the Mathematics Research Center, United States Army. It was held at the Wisconsin Center on the campus of the University of Wisconsin May 6-8, 1968.

The symposium was opened with words of welcome by Professor William H. Sewell, Chancellor, The University of Wisconsin and by Professor Barkley J. Rosser, Director of the Mathematics Research Center. The program consisted of five sessions in which altogether eleven papers were presented. The sessions were presided over by the following chairmen

Professor W. W. Peterson, University of Hawaii
Professor A. Hocquenghem, Conservatoire National des Arts and Métiers, Paris, France
Dr. David Slepian, Bell Telephone Laboratories, Murray Hill, New Jersey
Professor E. F. Moore, The University of Wisconsin
Professor D. K. Ray-Chaudhuri, The Ohio State University

The papers were concerned with research in algebraic coding theory and related areas of algebra and combinatorial theory.

The symposium committee consisted of the editor as chairman, Professors E. F. Moore and A. B. Fontaine both of the University of Wisconsin and Mrs. Gladys Moran whose services as a secretary greatly contributed to the success of the symposium. The manuscripts were prepared by Mrs. Doris Whitmore.

I take the opportunity to thank the speakers and chairmen and all others who contributed to this symposium.

Henry B. Mann

PREFACE

The last ten years have seen great advances in the study of algebraic codes and in the use of algebraic and combinatorial structures for their construction. The invention of the Bose-Chaudhuri codes, the finite geometry codes and the convolutional codes has given rise to many interesting problems in applications as well as in the arithmetic and algebra of finite fields and their polynomial domains.

The number of important contributions to this area is large and it was not easy to make a selection of topics and speakers. It was thought however that it would be best not to overload the program and to present only a small selection of papers from a fairly narrowly defined area. In this way sufficient time for informal discussion and interchange of ideas could be provided.

CONTENTS

ERROR CORRECTING CODES

F.J. MacWILLIAMS

An Historical Survey

I would like to start by expressing my gratitude to Miss Booth, of the Drafting Department at Murray Hill, for the beautiful chart.

To define the subject of this talk I shall quote an authority, Marcel Golay, as follows:

"The upper bound given by Shannon to the transmission capacity of a noisy discrete channel has challenged the mathematicians, who have accepted the challenge, to devise digital error correcting codes or coding systems approximating as close as possible to this upper bound."

I think we all agree with the definition contained in Golay's statement, and all disagree with what he says about mathematicians. Very few mathematicians have even noticed the challenge, let alone accepted it, for reasons which are well stated by another authority, Dr. Schützenberger, whom I quote

"The mathematics involved, even if quite simple, are far away from classical analysis, and indeed many of the necessary tools had to be sharpened especially for the purpose."

Most of the work in this field has been done by non-mathematicians, mainly by engineers; I suppose they are better at sharpening tools.

Since I have never been anything but a mathematician I am very poorly qualified to give this talk, particularly to an audience consisting largely of the engineers who did the work. I have not been able to suppress my mathematical prejudices — indeed, since I am here to amuse rather than instruct I didn't really try. Please don't start throwing things until the discussion period.

Our history, of course, begins with Shannon. However, in the pre-Shannon era there was a fairly abundant growth of primeval flora and fauna, resulting in deposits of valuable fuel which we are now beginning to discover.

After Shannon, the first problem attacked, and attacked first by Golay, is what I have called the Packing Problem, designated on the

Permutation groups 

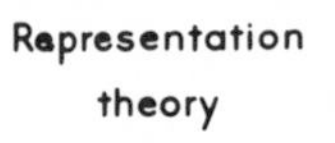
Representation theory 

1948 **Shannon**

	PACKING PROBLEM	LINEAR CODES		
			CYCLIC CODES	
				BCH CODES
1949	Golay Hamming			
1954	Plotkin		Reed-Muller codes	
1956		Slepian		
1957				
1958	Mathieu groups	*	Prange	
1959		* *	Fire codes	Hocquenghem Hamming codes cyclic
1960		Projective geometry	Reed-Solomon	Bose-Chaudhuri Peterson
1961			Group characters Fire codes cyclic	Number of information symbols 2-error Peterson
1962	Non-linear packings	*	Geometric codes π	
1963		Error location Coset codes	π	
1964	Rook domains	Error location Graph theory Tensor codes	Majority logic Pseudo-cyclic π Quadratic	Chien residues
1965		Coset codes Graph theory Tensor codes	Product codes Cyclic group algebras Punctured codes Pseudo cyclic	R M codes cyclic Symmetries Forney
1966	Non-linear codes	Symmetries	Difference set codes Shortened codes Synchronizing Dual product	Berlekamp Massey
1967		Search procedures Graph theory	Factored codes Geometric codes Self orthogonal Quasi cyclic	Number of information symbols
1968	Quadratic packings	Good non-cyclic codes	Dimension of geometric codes Product codes	G M R codes

Factorization of polynomials

M.I.T. CODES	LINEAR RECURRING (PSEUDO-RANDOM) SEQUENCES	OTHER "CODES"	REAL WORLD
	Error correction	Semi-groups	
	Arithmetic codes	Comma-free codes	
Recurrent codes	Polynomial rings Hamming codes	Variable length encoding	
$	Error correction	Prefix codes	Analysis of results
Sequential decoding Tree codes		Composition series	Detection and retransmission
Low-density codes	Composition series	Comma-free codes Linear residues	Prediction of error rates
Convolution codes Threshold decoding Analysis of $	Asynchronous multiplexing	Comma-free codes	Experimental retransmission
Recurrent codes	Error correction	Error distributing Linear residues	Experimental decoders Interleaving error rates
Convolution codes Burst correction Compare with block codes Military decoders	Fibonacci codes	Comma-free codes	Error rates Atmospheric noise detection
Concatenated codes Burst correction Error propagation Convolution codes	Asynchronous multiplexing	Prefix comma-free codes	Optimal block Digit error rates
Error bounds Tree codes Bounds on computation Error propagation	Shift register sequences Spread spectrum multiplexing		Military decoding Interleaving
	Cross correlations Arithmetic codes		Commercial data sets? Space decoders?

chart by a honeycomb. F is a finite field GF(q), F^n the vector space of n-place vectors over F. The distance (Hamming distance) between two vectors is the number of places in which they differ. The weight of a vector is its distance from the zero vector — the number of non-zero coordinates. To a vector S of F^n we attach a domain consisting of all vectors which differ from it in 1, 2, ..., e places; I shall call this by the picturesque name proposed by Dr. Golomb, a super-rook domain. The problem is to find as many as possible super rooks with non-overlapping domains in F^n — equivalently the distance between each pair of super rooks is to be at least $2e + 1$. The super rooks will be the messages which are sent over the noisy channel, and we have chosen the definition of distance between two vectors in such a way that vectors which are far apart are less likely to be mistaken for each other, when corrupted by noise, than vectors which are close together. At least one usually claims this to be the case.

If the non-overlapping domains exactly fill the space F^n we have a perfect e-packing or a closepacked code. In 1949 Marcel Golay discovered essentially all the perfect e packings that are known today, in spite of considerable efforts. About half the bees in the first column of the chart are looking for perfect e-packings or trying to prove that they do not exist, with rather meager results and considerable redundancy. I mean that the same paper is written and published several times. I will give an example.

If the vectors of a perfect e-covering form a linear subspace of F^n, we may define the dual space, that is the set of vectors whose scalar product with every vector of the covering is zero. This dual space also has distinctive properties; in particular, the dual of a perfect 1-packing (a Hamming code) consists of vectors of constant weight, except for the zero vector, of course. This space was discovered by Dr. Zaremba in 1952, in connection with properties of elementary abelian groups, and rediscovered about once a year from then on. I discovered it in 1960 and was very chagrined to find a long list of predecessors.

Let me point out a few queen bees. In 1957 Lowell Paige noticed that the Golay (23, 12) code can be arranged to be invariant under the permutations of the Mathieu group. This raises an interesting mathematical question; what is the largest permutation group which preserves a given code?

If we know the group, what can we say about the code, and if we know the code what can we say about the group? This is undoubtedly a difficult question, but it may sometimes have a simple answer. I was very interested to read that it has been shown that the group of an extended quadratic residue code is always simple — in a different sense of course.

In 1958 S. P. Lloyd determined a set of numerical conditions which are necessary (not sufficient) for the existence of a perfect e-packing. These were rediscovered by several people, including me. However, I was not as far behind as usual this time, I knew about Lloyd's result, and was trying to obtain a stronger one by imposing the condition that the perfect e-packing should be a linear space. In fact the numerical conditions came out the same — all I managed to do was to strengthen the hypothesis without weaking the conclusions. This of course gave rise to the conjecture that a perfect e-packing had to be a translation of a linear space. This is not true, it was disproved in 1962 by the discovery of a class of non-linear perfect 1-packings. All the same, there is something still to be elucidated. The strongest of Lloyd's numerical conditions is that the zeros of a certain rather horrible polynomial of degree e should be positive integers. If the perfect packing is a linear space, these integers are the weights of the vectors of the dual space. I would like to know what they represent otherwise.

We give up the idea of a perfect packing, but would still like it to fit as tightly as possible. The relation between the parameters e, n, q and the maximum number of non-overlapping rook domains is a problem which has no simple answers. Nonetheless, an enormous amount has been written about it — this is particularly odd since Hamming and Plotkin between them very early said practically everything which has been said since. About half the bees in the first column of the chart represent papers on this topic; only about five of them are innovations or real improvements on the work of Plotkin.

The Hamming distance is not the only possible metric, and indeed may not be the most appropriate. One of these bees is C. Y. Lee who pointed out that similar packing problems exist for other metrics, and similar results can be obtained. I believe that we shall hear more about this from Dr. Golomb.

The packing problem is a small part of Coding Theory, although represented by a lot of papers. Let us take a short step towards reality, and observe that at the receiving end of a noisy channel we have to decide whether or not the received vector is a super rook, and if not how it should be interpreted.

For some time, probably even pre-Shannon, computers were designed to use a parity check bit. The first computer of my acquaintance was so designed; if the word it was digesting contained an odd number of ones it stopped. This computer was not transistorized, it was not even vacuum-tubed, it operated with genuine relays. It had to be left running all night to get the roots of a fifth degree equation, and in the morning one would find it had stopped with a parity check failure. It is my belief that Hamming codes were invented or rediscovered by Dick Hamming because he got so aggravated with this computer. If the transistor had not become reliable we might have Hamming codes on computers today. However, this is beside the point; the point I wish to make is that the addition of several "parity checks" is a natural engineering extension of a system which was already in operation.

Now a set of vectors consisting of information symbols and linear parity checks is exactly equivalent to a linear subspace of F^n. A vector of the subspace is expressed in terms of a fixed basis of F^n, and the parity check equations are the relations of linear dependence between the coordinates of the vector. Moreover, if the parity check equations are applied to an arbitrary vector of F^n they tell us to which translation of the subspace the vector belongs. This became perfectly obvious as soon as Dave Slepian pointed it out in 1961. The effect of his paper, at the time it was written, was like that of turning on a searchlight.

The general problem in the theory of linear codes is to describe the properties of a subspace of F^n which depend on a fixed basis of F^n. This is the antithesis of classical algebra and very frustrating to classical mathematicians. The papers which I consider to be attacks on the general problem are disappointingly few, and many of them are simply restatements of the problem in a more abstract and, to me, less useful form. I recall one which said that the Grothendieck group of the category of linear codes is a λ-ring under the operation of tensor product of codes, and that this is Dave's theorem so-and-so. When I asked Dave about this he indignantly denied ever having said any such thing.

Leaving the general problem almost untouched, we pass to the subdivision cyclic codes. These are linear codes with the added restriction that if $(a_0, a_1, \ldots, a_{n-1})$ belongs to A so does $(a_{n-1}, a_0, a_1, \ldots, a_{n-2})$. In fact we have introduced a multiplication of vectors; the code is an ideal in the polynomial ring $F[x]/x^n-1$. The problem is nearer to classical mathematics; however, perhaps more important is the fact that it can be dealt with in terms of shift registers and linear recurring sequences. These are engineer's tools, and most of these papers are by engineers.

Cyclic codes as such were introduced by Eugene Prange in 1959 or '60, although there are earlier papers which use them implicitly. The big rush started with the Bose-Chaudhuri-Hocquenghem theorem, which provides a lower bound for the weight of vectors of a cyclic code. It was very quickly discovered that almost every special code previously invented could be arranged to be cyclic. Hamming codes are cyclic, Golay codes are cyclic, Reed-Muller codes are cyclic and so on.

The symbols on the chart represent papers which I found in a not very exhaustive search of the literature — that is I was exhausted long before the literature was. After 1962 they are certainly only a fraction of what was actually published.

As I understand it, BCH codes are a subclass of cyclic codes — a BCH code in the wide sense is the largest code which has a certain minimum distance guaranteed by the BCH bound. I have tried to make a distinction on the chart between papers on cyclic codes in general and papers on BCH codes, but I was considerably hampered by the fact that many authors put the magic initials BCH in the title regardless of the content of the paper. One example of a horrible misnomer is this paper, which is called "A New Look at BCH Codes." The real point made by the authors is that the character theory of a cyclic group algebra is a useful tool for the study of cyclic codes in general; and so it has proved, the paper has many successors. Personally I don't think it necessary, or even advisable, to derive character theory from linear recurring sequences, but I am bound to admit that no one else had thought of using it at all.

In contrast to the previous situation, most of these papers represent genuine contributions, many more than I can possible describe.

First we have Dr. Peterson's fundamental decoding algorithm for BCH codes. (The spectacles stand for decoding.) This was embellished and elaborated by many of these bespectacled gentlemen, with the aim of obtaining less costly logical implementations.

A second decoding method, permutation decoding, was discovered by Prange, and again elaborated by several other authors. (π + spectacles is permutation decoding.) The idea is to find a set of permutations which leave the code invariant, and such that all correctable error patterns are transferred to the parity check places by at least one permutation of the set. The error can then be easily identified.

A third method, majority logic decoding, was proposed by Luther Rudolph in 1962. This depends on finding a set of parity check equations which are orthogonal in the Massey sense, that is they all check one particular information bit and are otherwise disjoint; one then takes a vote as to the correctness or not of the particular information place. Surprisingly, Rudolph's work was completely ignored until Ned Weldon rediscovered the idea independently in 1966.

It is, of course, of practical importance to know the number of information symbols in the code one is proposing to decode. A method of finding this for BCH codes was discovered quite early by Dr. Mann, and lately extended by Elwyn Berlekamp. The same problem for the geometric codes used in majority logic decoding has just been solved by Dr. Goethals.

All of these decoding methods demand quite a lot of not so simple mathematics. Peterson decoding depends on factoring polynomials over finite fields; permutation decoding in finding a suitable permutation group, and majority logic decoding, at least at present, on the properties of finite geometries. I have learned, by attending this conference, that topological codes are also decoded by majority logic, and I hope someday to find out whether there is a connection between simplexes and finite geometries. In all cases, excavations in the pre-Shannon era have proved very useful, although again a number of new tools have had to be forged and sharpened.

The squarish object is supposed to be a weight, and represents a paper on the weight structure of a particular code or class of codes. So much of this work has been done at the University of Hawaii that I almost made it a pineapple instead. The stylized explosion stands for a paper on correction of bursts of errors, a very popular subject and one of great practical importance. The snail is a paper I did not wish to classify; it is not intended as an insult — in fact it was not intended to be a snail. I drew a sort of circular squiggle, and the draftsman, or rather draftswoman, Miss Booth, who was copying the chart for me, thought I meant it to be a snail. The snail that she drew is so pretty that I decided to keep it. It stands for a paper on a topic which does not occur very frequently — for example, conditions under which BCH code does or does not attain its BCH bound. The fleur-de-lis stands for a paper in French, of which there are more than I have indicated. I apologize to the Russians who have also made a sizeable contribution to the subject. I could not think of a symbol for Russian which was easy to draw.

We now interrupt this program for the commercial. Elwyn Berlekamp has written a book, about to be published by McGraw-Hill, which I can only describe as an encyclopedia of cyclic codes. It also contains a very extensive bibliography, of which I made use in preparing this talk. I cannot produce an animated cartoon at this point, but I have composed an advertising lemma-ric [1] which I will now recite.

> The Berlekamp book you must get
> To belong to the cyclic code set.
> It purports to state
> All knowledge to date
> And a lot that is not known as yet.

Be sure that your library has at least two copies.

I have called the next section M.I.T. codes because most of the work listed there has been done at M.I.T. or by people from M.I.T. and is well documented in books published by the M.I.T. press. The previous sections are very poorly documented — Elwyn's book will be only the second.

Low Density Parity Check Codes and Concatenated Codes are linear codes in the sense of Slepian, although, of course, of a very special kind. Essentially everything that is known about them is contained in books by their inventors, Dr. Gallagher and Dr. Forney respectively.

In recurrent or convolutional codes the parity check bits are inserted between the information bits by a set of sliding parity check equations. The whole message is one vector of an enormous linear code — a fact which is no help at all. The decoding is done one bit at a time, using the previously received and corrected bits, so that the decision for the current bit depends on the decisions for an unknown number of previous bits. (Unknown to me that is, because I am not sure how to interpret the phrase "depends on" in the previous sentence.) Ideally the whole message is stored and worked over until the decoder is satisfied that it has made the most probable decision, taking into account the statistics of the channel. Sequential decoding, introduced by Wozencraft and Reiffen, tries to approximate this ideal situation. The dollar sign represents an experimental sequential encoder and decoder built and operated by Dr. Wozencraft, using a "random" convolutional code — "random" means that he did not try very hard to optimize it. I don't know what a sequential decoder looks like, so I use a dollar sign to indicate the scope and thoroughness of this experiment. The practical problem turned out to be, as predicted by the theory, that occasionally the decoder spends a very long time working on a particular bit, even though the average time is quite small. There are a number of recent papers on this problem, and as a matter of fact a number of Russian papers.

In the other type of decoding one gets rid of the corrected bit as soon as it is no longer involved in the parity checks in the current bit. If the correction is fed back into the decoder it is called feedback decoding, if not it is called direct-decoding; this is a confusing choice of names since direct decoding is in many ways the less direct scheme. In both cases it is necessary to work on optimizing the code and the problem is a difficult one which is not amenable to standard mathematical or engineering techniques. Nonetheless, we do know a certain amount, and we shall hear about this from Dr. Massey himself. I should add that real encoders and decoders using this scheme have been built for military use.

The next two columns of the chart describe work which is perhaps not strictly error correction by Golay's definition, although you see there a number of papers on error correction or burst error correction. These are really concerned with the properties of maximal length shift register sequences, which I prefer to regard as minimal ideals in the polynomial ring $F[x]/x^n - 1$, where $n = q^m - 1$. They really belong in the section on cyclic codes. There are also fascinating problems connected with the cross-correlation of a sequence with itself, that is the scalar product of a vector with its cyclic permutations. Sequences for which this is small have some very practical applications, but I cannot claim that they are really error-correction. I have also put arithmetic codes in this column; these are designed to check machine errors in computation, and are now fairly well understood.

We come to what I have called the Real World. I should point out that my experience with the Real World is definitely limited, being confined to what went on in the next two corridors at the Holmdel laboratories during the time that I worked there. Moreover, there are people here who still inhabit these corridors, and are much better qualified than I am to talk about their work.

The woodpecker in the telephone pole represents a series of observations of data transmission over the telephone network. These observations — that is the actual locations of the actual errors in a large number of trials — were used to test various error control schemes both by computer simulation and statistical analysis. It was immediately apparent that random error correction with a medium length ($n < 100$) BCH code was worse than no code at all. The bit error rate went up when the code was used. The code detected the errors all right, but it corrected them all wrong. For a while the favorite scheme was detection and retransmission, which theoretically looked very attractive. An experimental encoder and detector for a (60, 48) shortened cyclic code was built, and ran for several hundred hours with absolutely no undetected errors. But there was a difficulty — the same trouble as with sequential decoding. The incoming data, held up by requests for retransmission, caused buffer overflows fairly frequently, in fact every two or three hours.

The next experiment was with a (15, 9) cyclic code used for burst correction, and also interleaved to depth 200 . That is 200 code words are collected, the first bit of each word is transmitted, then the second bit of each word and so on. This experiment was successful in one sense and unsuccessful in another. The number of bursts which got through was very much reduced, in fact by a factor of something like 200; however, the bit error rate was reduced only by a factor of 14 . This, of course, was because a burst that got through was a real disaster.

I have heard a rumor that IBM, has built, for military use, an encoder and decoder using the Golay (23, 12) code, interleaved to a large degree and at a price that only the military could afford. So we jump from the top left to the bottom right hand corner of the chart, with a couple of stops on the way to pick up information on how to encode and decode the Golay code.

The list of applications is decidedly meager when compared with the amount of theory, and I have the feeling that someone ought to be worried. It is my heretical opinion that we have put too much effort into the study of cyclic codes. We have neglected other areas which may be of more practical importance for error correction in a real situation.

If what we need to cope with the telephone network or other real life channels is a powerful code of enormous block length we should start looking for it, even if the mathematics is not as elegant as it is for BCH codes. The alternative [2] of scrapping the telephone network in favor of a system better adapted to random error correction is not really practical — although I agree that new facilities should be designed with data transmission in mind. Dr. Massey has described my feelings very well in another lemma-ric, and I conclude by reciting his composition.

Delight in your algebra dressy,
But take heed from a lady named Jessie
Who spoke to us here
Of her primitive fear
That good codes just might be messy.

REFERENCES

[1] This pun was committed by Dr. S. Golomb.

[2] Proposed by Dr. E. C. Posner.

EDWARD C. POSNER

Combinatorial Structures in Planetary Reconnaissance

Summary

This paper describes those portions of the Mariner '69 High-Rate Telemetry System of interest to combinatorial mathematicians. After describing the Mariner '69 High Rate System, we list the gains in performance attributable to various factors including coding. We then explain the reasoning behind the choice of the particular biorthogonal coding scheme used. Encoding and decoding are then described; decoding is shown to be a special case of the Fast Fourier Transform. The method of word synchronization using a coset of the original biorthogonal Reed-Muller code is then presented. We then describe the receiving equipment already built, and close the paper with a list of some areas of combinatorial research suggested by the Mariner '69 High-Rate System.

1. Introduction

In February or March 1969, the Mariner '69 spacecraft is due to be launched toward the planet Mars, with encounter scheduled for about 5 months later. The prime purpose of the mission is to return high-quality photographs of the planet. To this end, a high-rate digital picture transmission system has been designed to operate at 16.2 kilobits/sec. Since the Mariner IV Mars mission had rate of only 8-1/3 bits/sec., it is interesting to see how the 32.9 db gain arose. Table I compares the Mariner IV system with the Mariner '69 High-Rate System. Of mathematical interest is the improved symbol synchronization system on Mariner '69, which saves 5.2 db. However, it is not that aspect that concerns us here. Instead, it is the biorthogonal coding scheme that saves 2.2 db which is the concern of this paper.

The chosen code is a coset of the (32, 6) biorthogonal Reed-Muller code, comma-free of index 6. The hard problem is to decode at the high information rate. The decoding scheme chosen turns out to be a Fast Fourier Transform on a direct sum of five groups of order 2;

TABLE I. COMPARISON OF TELEMETRY SYSTEMS

PARAMETER	MARINER IV	MARINER '69 HIGH-RATE	IMPROVEMENT
Transmitter Power	(8.9W) + 39.5 dbm W	(18.2W) + 42.60 dbm W	*+3.10 db
Modulation Loss	-5.3 db	-1.34 db	*+3.96 db
Transmitting Antenna Gain	+20.1 db	+20.21 db	*+.11 db
Space Loss	(216×10^6 km) -266.2 db	(97×10^6 km) -259.36 db	*+6.84 db
Receiving Antenna Gain	(85 ft.) +52.5 db	(210 ft.) +61.00 db	*+8.50 db
Receiver Loss	-3.4 db	-.44 db	*+2.96 db
Received Sideband Power	-162.8 dbm W	-137.33 db	+25.47 db
Bit Rate	(8-1/3 bit/sec) -9.2 db/s	(16.2×10^3 bit/sec) -42.10 db/s	-32.90 db
Bit Energy	-172.0 dbm Ws	-179.43 dbm Ws	-7.43 db
Receiver Noise Temperature	(65°K) -180.5 dbm W/Hz	(25°K) -184.60 dbm W/Hz	*+4.10 db
Nominal Signal-to-Noise Ratio Per Bit	+8.5 db	+5.17 db	-3.33 db
Signal-to-Noise Ratio Required for Given Bit Error Probability	(5×10^{-3}) +5.2 db	(5×10^{-3}) +3.00 db	*+2.20 db
Margin	+3.3 db	+2.17 db	*-1.13 db
Total Gain in Bit Rate (Sum of starred items in last column)	--------------------	----------------------	+32.90 db

it is this possibility of fast decoding that led to the choice of a Reed-Muller code.

In Section 2, we give some of the reasons which led to the particular design of the coding system used. Section 3 describes encoding, and Section 4 decoding. Section 5 describes the word synchronization system. In Section 6, a description of equipment already demonstrated is given. Section 7 closes the paper with some combinatorial problems whose solution would be useful to designs of block-coded self-synchronizing telemetry systems of the kind used in the Mariner '69 High-Rate System.

Let us close this section with an acknowledgement to some of the many engineers and mathematicians instrumental in the design of the Mariner '69 High-Rate System, and in the preparation of this paper. Thanks are due to Tage O. Anderson, Leonard D. Baumert, Mahlon Easterling, Solomon W. Golomb, Richard Green, Norman De Groot, Murray Koerner, William C. Lindsey, Howard Rumsey Jr., Paul Schottler, Robert C. Tausworthe, Andrew J. Viterbi, Lloyd Welch, and Robin Winkelstein.

2. Choice of Biorthogonal Code

What was desired was a digital telemetry system operating at 16.2 kilobits per second, with a bit error rate of at most 5×10^{-3}. As Table I shows, this error probability would not have been achievable without coding. A higher bit rate is desirable so as to reduce the probability of spacecraft or tracking station failure during reception, and to maintain the favorable pointing angle of the spacecraft high-gain antenna which is achieved at encounter. Conditions of bandwidth in the receiver prevented an expansion of the bandwidth of much more than the 5-1/3 to 1 achieved with the (32, 6) code.

A biorthogonal code was chosen rather than an orthogonal code because a (32, 6) biorthogonal code is as easy to implement as a (32, 5) orthogonal code, since the correlation values of the received waveform with the (32, 6) code consist of the correlations with the (32, 5) code together with their negatives. And the (32, 6) code yields .08 db more than the (32, 5). A (64, 6) orthogonal code would take twice as much correlation equipment and bandwidth, and still would provide fewer db than a (32, 6) biorthogonal code.

More gain (less than .1 db, however,) would have been provided with a (63, 6) transorthogonal code, the optimum for 6 bits/word in a Gaussian channel. But the bandwidth expansion would have been too high, and the correlating equipment twice as complex. A (31, 5) transorthogonal code, on the other hand, is actually a little worse (by less than .1 db) then a (32, 6) biorthogonal code; see [1] for these matters. Thus, a (32, 6) biorthogonal code was adopted.

Exactly which (32, 6) biorthogonal code to use was dictated by decoding speed with fixed equipment. The so-called Green Machine

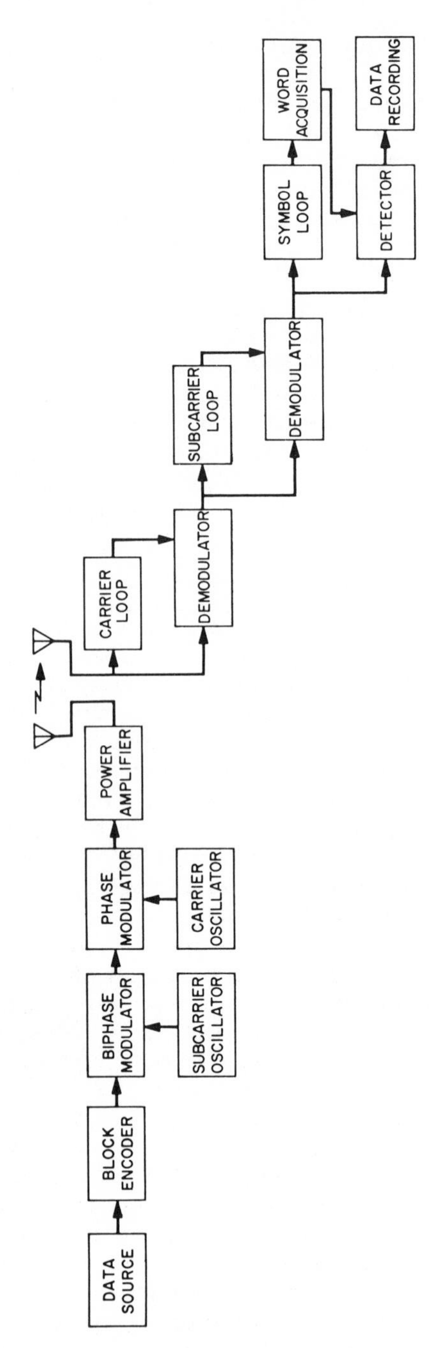

Fig. 1. High-rate telemetry system block diagram

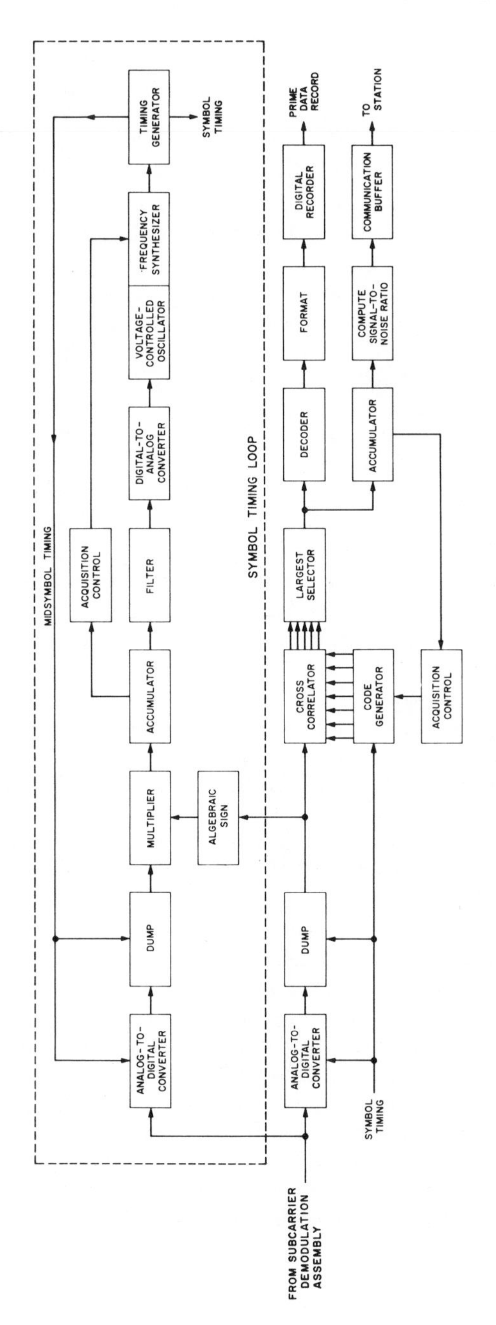

Fig. 2. High-rate ground telemetry block diagram

which correlates the received telemetry waveform with an entire (32, 5) orthogonal code dictionary uses a version of the Fast Fourier Transform for finite abelian groups, and is available only for the so-called Reed-Muller codes. Otherwise the correlations could be performed at only $\frac{5}{32}$ the rate achieved with the Green Machine.

Since a block code was adopted, some method of achieving word synchronization was needed. The simplest method, from the point of view of the spacecraft, is to let the code be self-synchronizing. This was done by using a coset of the (32, 6) biorthogonal Reed-Muller code rather than the code itself, as we shall see in Section 5. The particular choice of coset was dictated more by simplicity of generating the coset on board the spacecraft, rather than by trying for the largest possible index of comma-freedom for any coset of the (32, 6) code.

Figure I gives a block diagram of the Mariner '69 High-Rate Telemetry System, including both the spacecraft and ground portions. Figure II is a more detailed block diagram of the ground portions of the High-Rate System. The analogue-to-digital converter in the Crosscorrelator Assembly of Figure II quantizes the integrated symbols into 13 bits, so that the actual correlations are computed digitally from these values; 18 bits are necessary for the final values. This represents only an infinitesimal loss over full analogue correlation; even 3 bit quantization would lose less than .1 db .

It should be pointed out [2] that 1 bit quantization, i.e., sybol-by-symbol detection, loses so much (more than .9 db if word error probability is the criterion, even more for a bit error criterion) as to be avoided.

For the sake of completeness, we give a brief description of the radio portion of the Mariner '69 High-Rate Telemetry System. The carrier frequency is 2295 mHz; a square wave subcarrier of frequency 259.2 kHz is used, which is 3 times the symbol rate of $(32/6) \times 16.2 = 86.4$ kilobits/sec. The modulation is biphase with a modulation index of 13/18, corresponding to 65° modulation.

We are now ready to study the combinatorial aspects of the Mariner '69 High-Rate Telemetry System. These aspects occur in the code structure, in the encoding and decoding, and in the word synchronization. These are quite interesting aspects, but the mathematical reader should remember the vast effort in highly precise, difficult, and imaginative engineering that went into the system. After all, coding provides only 2.2 db of the 32.9 db gain in information rate of the Mariner '69 system over the Mariner IV system!

3. Encoding

This section describes the Mariner '69 (32, 6) biorthogonal Reed-Muller code and its encoding. First the (32, 5) orthogonal

Reed-Muller code will be described. Consider the Hadamard matrix

$$H_1 = \begin{pmatrix} 1 & 1 \\ 1 & -1 \end{pmatrix} ,$$

and define the Hadamard matrix H_{n+1} by

$$H_{n+1} = H_n \otimes H_1, \quad n \geq 1 .$$

For example,

$$H_2 = \begin{pmatrix} 1 & 1 & 1 & 1 \\ 1 & -1 & 1 & -1 \\ 1 & 1 & -1 & -1 \\ 1 & -1 & -1 & 1 \end{pmatrix} .$$

The $(32, 5)$ Reed-Muller code used is H_5; the $(32, 6)$ biorthogonal Reed-Muller code consisting of

$$\begin{pmatrix} H_5 \\ -H_5 \end{pmatrix}$$

is shown in Figure III. The $(32, 5)$ code can also be though of as consisting of the first 32 Payley-Walsh Functions, if each column of the 32×32 Hadamard matrix is regarded as an interval of length $1/32$.

Now that the code set has been described, the correspondence between 6-tuples of information bits and codewords will be given. Let

$$\phi : \{1, -1\} \rightarrow \{0, 1\}$$

be the unique isomorphism of the multiplicative group of two elements onto the additive group of two elements, so that (obvious notation) $\phi(H_5)$ is a code over $\{0, 1\}$. The correspondence between five-tuples x of information bits and rows of $\phi(H_5)$ is then given as follows. Let

$$x = (x_4, x_3, x_2, x_1, x_0) ,$$

and let

$$\lambda(x) = \sum_{i=0}^{4} x_i 2^i \quad \text{(here } x_i \text{ is regarded as real)} .$$

Then, if the rows of H_5 are numbered from top to bottom as 0 to 31, the five-tuple x in the (32, 5) Reed-Muller code is encoded as the $\lambda(x)^{\underline{th}}$ row of $\phi(H_5)$.

The encoding procedure actually used on board the Mariner '69 spacecraft is the following. Consider as in Figure IV the 32×5 matrix K_5 with rows numbered from 0 to 31 whose $j^{\underline{th}}$ row is the five-tuple consisting of the binary expansion of j. Note that the 5×32 matrix

$$\phi^{-1}(K^T)$$

consists of the first 5 Rademacher functions, with a suitable definition of the unit interval. Encoding is simply the following: the five-tuple

$$x = (x_4, x_3, x_2, x_1, x_0)$$

is encoded as the modulo-two sum

$$\sum_{i=0}^{4} x_{4-i} k_i ,$$

where k_i is the $i^{\underline{th}}$ column of K_5^T, the columns being numbered from 0 to 4. In other words, the (32, 5) Reed-Muller code is a linear code, with basis vectors

$$\{k_{4-i}, \ 0 \le i \le 4\} .$$

The code is systematic, the information bits appearing "in the clear" as symbols 1, 2, 4, 8, 16.

It must be shown that this encoding procedure does in fact lead to the same code as does the Hadamard matrix definition, with the same correspondence between information bits and codewords. Let

$$K_1 = \begin{pmatrix} 0 \\ 1 \end{pmatrix} ,$$

and, for $n \ge 1$, define

$$K_{n+1} = \begin{pmatrix} 0, & K_n \\ 1, & K_n \end{pmatrix} ,$$

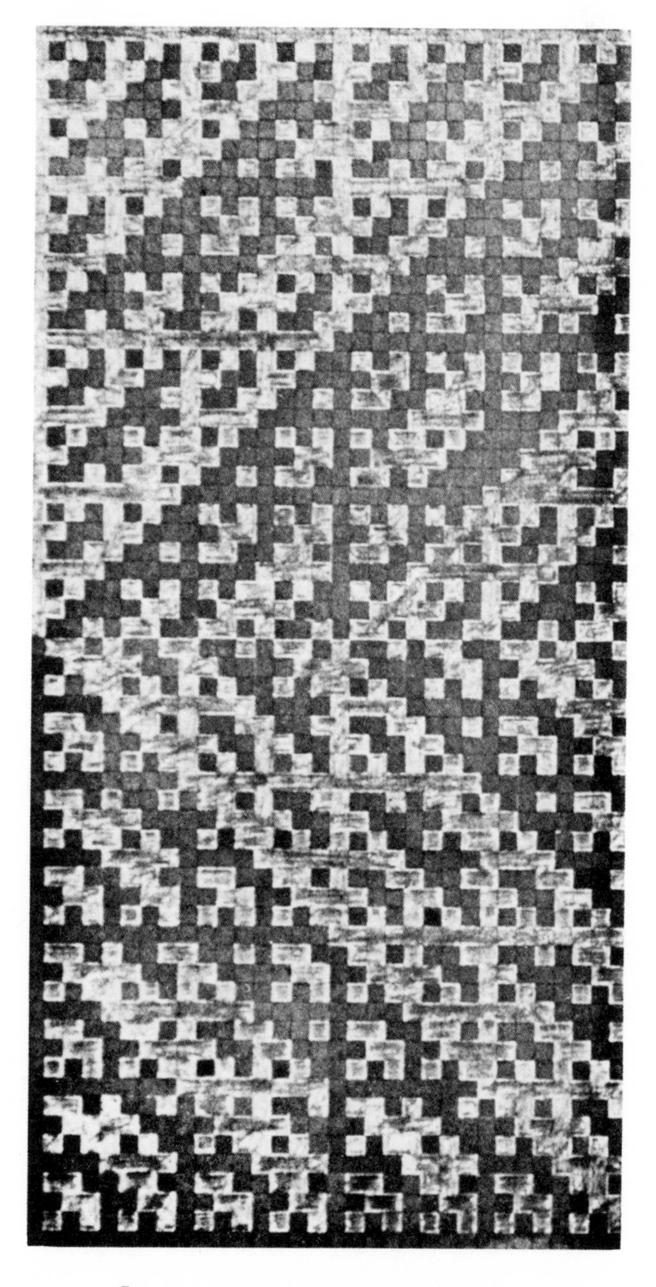

Fig. III. The (32, 6) Biorthogonal Reed-Muller Code

FIGURE IV

The Matrix K_5

0	0	0	0	0
0	0	0	0	1
0	0	0	1	0
0	0	0	1	1
0	0	1	0	0
0	0	1	0	1
0	0	1	1	0
0	0	1	1	1
0	1	0	0	0
0	1	0	0	1
0	1	0	1	0
0	1	0	1	1
0	1	1	0	0
0	1	1	0	1
0	1	1	1	0
0	1	1	1	1
1	0	0	0	0
1	0	0	0	1
1	0	0	1	0
1	0	0	1	1
1	0	1	0	0
1	0	1	0	1
1	0	1	1	0
1	0	1	1	1
1	1	0	0	0
1	1	0	0	1
1	1	0	1	0
1	1	0	1	1
1	1	1	0	0
1	1	1	0	1
1	1	1	1	0
1	1	1	1	1

where "0" and "1" denote 2^n-tuples of 0's and 1's in lexicographic order. The encoding procedure first described leads to the code

$$K_n K_n^T ,$$

where the ordering of the rows is lexicographic on the binary expansion of the index of the row. We are to prove that

$$K_n K_n^T = \phi(H_n), \ n \geq 1 .$$

When $n = 1$,

$$K_1 K_1^T = \begin{pmatrix} 0 & 0 \\ 0 & 1 \end{pmatrix} = \phi(H_1) .$$

Assuming the result for n, we prove it for $n+1$ as follows. Write

$$K_{n+1} K_{n+1}^T = \begin{pmatrix} 0, & K_n \\ 1, & K_n \end{pmatrix} \begin{pmatrix} 0^T, & 1^T \\ K_n^T, & K_n^T \end{pmatrix}$$

$$= \begin{pmatrix} K_n K_n^T & K_n K_n^T \\ K_n K_n^T & 11^T + K_n K_n^T \end{pmatrix} .$$

Since 11^T is the $2^n \times 2^n$ matrix all of whose entries are equal to 1, we conclude

$$\phi^{-1}(K_{n+1} K_{n+1}^T) = \phi^{-1}(K_n K_n^T) \otimes H_1$$

$$= H_n \otimes H_1 = H_{n+1} ,$$

as promised.

The actual mechanization of the encoder is shown in Figure V. A binary counter of 5 stages counts from 0 to 31, least significant bit right-most; the information 5-tuple x acts on the counter at time j to produce the $j^{\underline{th}}$ symbol w_j of the encoded word w by the formula

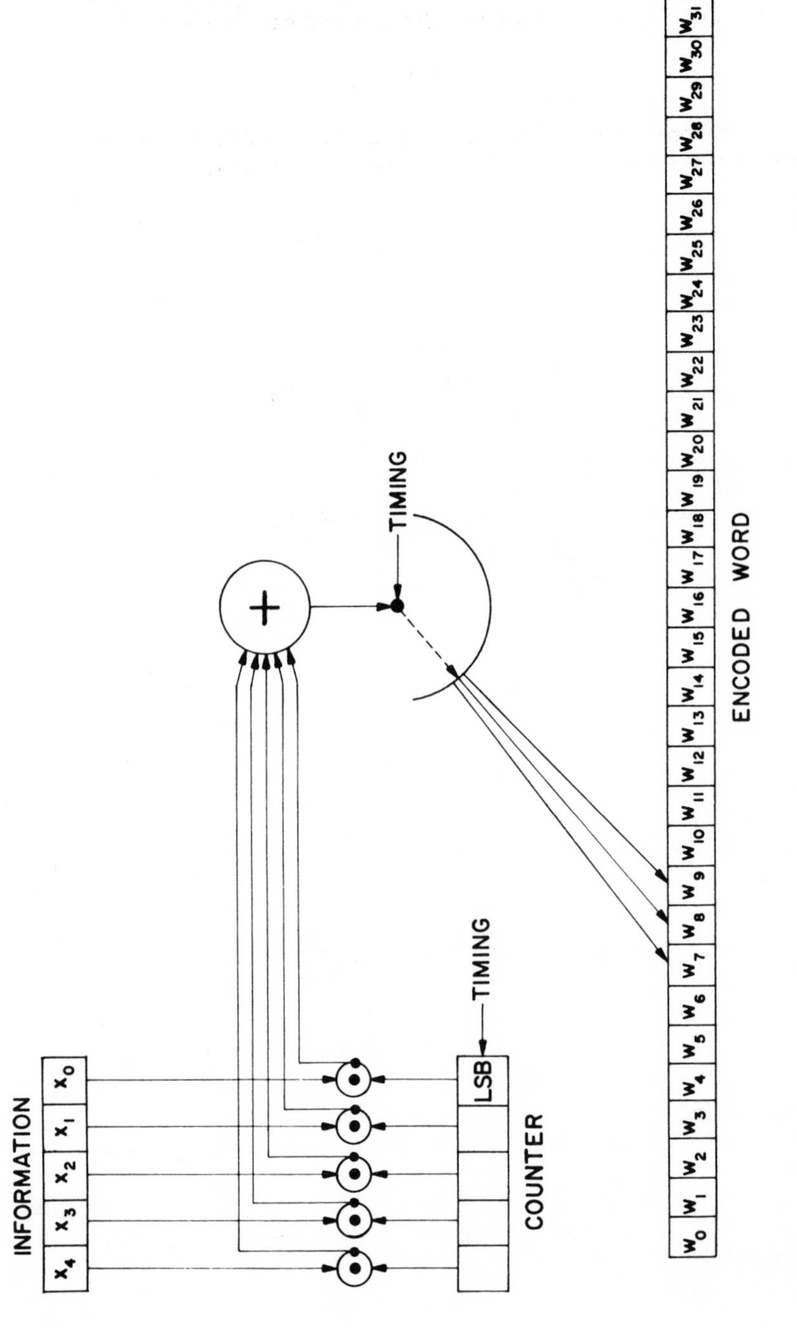

FIGURE V. THE ENCODER FOR THE (32,5) REED-MULLER CODE

$$w_j = \sum_{\ell=0}^{4} j_\ell x_\ell ,$$

the sum being modulo 2, where

$$j = \sum_{\ell=0}^{4} j_\ell 2^\ell .$$

The encloded word w is then

$$w(x) = (w_0, w_1, \ldots, w_{30}, w_{31}) .$$

The sixth information bit x_5 is encoded as follows: If x_5 is a 0, transmit $w(x)$ where

$$x = (x_4, x_3, x_2, x_1, x_0) .$$

If x_5 is a 1, transmit

$$\bar{w} = (1+w_0, 1+w_1, \ldots, 1+w_{31}) .$$

This assignment minimizes the bit error probability for given word error probability, since opposite information 6-tuples are encoded as opposite 32-tuples. The resulting $(32, 6)$ biorthogonal Reed-Muller code is then systematic on positions $0, 1, 2, 4, 8$, and 16 .

Another encoding procedure, discovered by T. O. Anderson, is essentially the dual of the decoding procedure to be given in the next section. It is n times as fast, with similar equipment, as the encoding procedure described above, thus 5 times as fast for Mariner '69. Although encoding speed was not a problem for this mission, there may be missions where the data rate is so high that the fast encoder would be useful.

4. Decoding

This section describes the decoding procedure used for the $(32, 6)$ biorthogonal Reed-Muller code in the Mariner '69 High-Rate Telemetry System; a procedure of this general nature was originally suggested by Murray Koerner of JPL. Conceptually, one first strips off the so-called comma-free vector that was added on to obtain a coset of the Reed-Muller code; no further mention will be made of this operation until the next section. Then one computes the correlations of the received telemetry waveform with each codeword of the (32, 5) Reed-Muller code, regarded as a code over $\{1, -1\}$. The codeword producing the correlation of largest absolute value is the maximum-likelihood estimator of the transmitted word if this correlation is positive; otherwise, the negative of the word is chosen.

If one generates the codewords of a $(2^n, n)$ Reed-Muller code sequentially and then correlates each time, 2^{2n+1} operations are required to perform the correlations: 2^n codewords with 2^{n+1} "operations" (real additions and multiplications). If a smaller work factor were attainable, less equipment would be necessary, less time would be needed, or both.

The method adopted in the Mariner '69 High-Rate Telemetry System has a work factor of only $n \cdot 2^{n+1}$, a saving of a factor of $2^n/n$ or 6.4. This saving is realized mainly in information rate. The method utilizes the Fast Fourier Transform as generalized to finite abelian group by Lloyd Welch [3], although this special case was discovered independently by Richard Green [4]. In fact, the possibility of using Fast Fourier Transform decoding is what makes the use of Reed-Muller codes so attractive.

The idea is this. Let us identify the word time with the unit interval. The received waveform averaged over the symbol period is identified with a function $f(g)$ on the set G of 32 symbol periods, as are the 2^n codewords of the Reed-Muller $(2^n, n)$ code. We make the set of 2^n symbol periods into a finite abelian group, also called G, by identifying the interval $[j/2^n, (j+1)/2^n)$, $0 \leq j \leq 2^n - 1$, with the n-tuple consisting of the binary expansion of j; addition is componentwise modulo 2. Then f is expanded into its finite Fourier Series on G as

$$f(g) = \sum_{\chi \in \hat{G}} \hat{f}(\chi)\, \chi(g) \quad ,$$

where $\hat{G}$ is the character group of G.

What is the character group of $\hat{G}$ of G? If $\chi \in \hat{G}$, there is a Payley-Walsh function w, one of the first 2^n Payley-Walsh functions, such that

$$\chi = \chi_w$$

where

$$\chi_w(g) = \int_g w(t)\, d(2^n t) \quad ;$$

here g under the integral denotes the symbol interval corresponding to g. It needs to be verified that

$$\chi_w(g_1+g_2) = \chi_w(g_1)\,\chi_w(g_2) \quad .$$

However, this fact is immediate for the n Rademacher functions (essentially the n columns of the matrix K_n of Sec. 3), and follows for the other Payley-Walsh functions by the multiplicativity of the correspondence

$$w \leftrightarrow \chi_w \quad .$$

The Fourier expansion of f then becomes

$$f(g) = \sum_w \hat{f}(\chi_w) \int_g w(t)\,d(2^n t) \quad .$$

The Fourier inversion formula gives

$$\hat{f}(\chi_w) = \sum_{g \in G} \frac{1}{2^n} f(g) \int_g w(t)\,d(2^n t) \quad ,$$

or

$$\hat{f}(\chi_w) = \int_{t=0}^{1} \underset{\sim}{f}(t)\,w(t)\,dt \quad ,$$

where

$$\underset{\sim}{f}(t) = f(g), \quad t \in g \quad ,$$

the symbol interval corresponding to g. In other words, $\hat{f}(\chi_w)$ is nothing but the correlation of received waveform f with the Payley-Walsh function w.

Now as we saw in Sec. 3, the set of the first 2^n Payley-Walsh functions is equal to the set of the 2^n rows of the Hadamard matrix H_n, if the 2^n columns indices of H_n are identified with the 2^n symbol intervals in the same order of left to right. Thus, to correlate a received waveform f with the $(2^n, n)$ orthogonal Reed-Muller code, it is enough to compute the Fourier Transform of f as a function on G. And whenever one thinks of Fourier Transform these days, one thinks of Fast Fourier Transform (FFT). Since the FFT on finite abelian groups is not well-known, a summary of it will be given first.

Let G be a finite abelian group with character group $\hat{G}$. It is desired to compute for f a say real-valued function on G the Fourier Transform

$$\hat{f}(\chi) = \sum_{g \in G} f(g)\chi(g), \quad \chi \in \hat{G} \quad .$$

Let the order of G, $|G|$, be written as

$$|G| = p_1 p_2 \cdots p_k \quad ,$$

p_i not necessarily distinct primes, and let

$$G = G^{(k)} > G^{(k-1)} > \ldots > G^{(1)} > \{e\} = G^{(0)}$$

be a composition series for G with

$$|G^{(j)}/G^{(j-1)}| = p_j, \; 1 \leq j \leq k \; .$$

Let g_j be in $G^{(j)}$ but not in $G^{(j-1)}$; every choice of composition series and g_j leads to a different FFT. Then every g in G is uniquely expressible as

$$g = \prod_{j=1}^{k} g_j^{\ell_j}, \quad 0 \leq \ell_j \leq p_j - 1 \; .$$

The Fourier Transform now becomes

$$\hat{f}(\chi) = \sum_{\ell_k=0}^{p_k-1} \cdots \sum_{\ell_1=0}^{p_1-1} f(g_1^{\ell_1} \cdots g_k^{\ell_k}) \chi(g_1^{\ell_1} \cdots g_k^{\ell_k}) \quad ,$$

or

$$\hat{f}(\chi) = \sum_{\ell_k=0}^{p_k-1} \chi(g_k^{\ell_k}) (\ldots (\sum_{\ell_1=0}^{p_1-1} \chi(g_1^{\ell_1}) \; f(g_1^{\ell_1} \cdots g_k^{\ell_k})) \ldots) \; .$$

Since the cyclic group $\{g_j\}$ generated by g_j has only p_i characters, not $|G|$ characters, we need evaluate the sum over ℓ_j for only p_j choices of χ; the evaluation of $\chi(g_j^0)$ requires no calculation.

Thus, at the $j^{\underline{th}}$ step, there are $p_1 p_2 \cdots p_j$ different choices of character, and $p_{j+1} \cdots p_k$ choices of the remaining exponents $\ell_{j+1}, \ldots, \ell_k$. That is, removing one pair of parentheses requires

$$|G|[2(p_j-1)]$$

calculations, namely, p_j-1 additions and p_j-1 multiplications for each of $|G|$ possibilities of character and exponent. Thus, to compute $\hat{f}(\chi)$ for all χ requires a work factor of

$$2|G| \sum_{j=1}^{k} (p_j - 1) \quad .$$

By contrast, computing $f(\chi)$ for each χ separately requires

$$2|G|(|G|-1)$$

operations. That is, for each of $|G|$ characters, $|G|-1$ additions and $|G|-1$ multiplications are required. The savings in work factor achieved by the FFT is therefore

$$\frac{|G|}{\sum_{j=1}^{k} (p_j - 1)} \quad .$$

In the case of interest for the Mariner '69 High-Rate System, $k = n$ and $p_j = 2$, $1 \leq j \leq n$, for a saving of a factor of $32/5$ when $n = 5$.

The ordinary (trigonometric) FFT for 2^n equally spaced points on the unit interval starting with 0 has for its G the additive group modulo 2^n, under the correspondence

$$j/2^n \rightarrow j \text{ modulo } 2^n, \; 0 \leq j \leq 2^n - 1 \quad .$$

The characters are

$$\chi^{(m)}(j) = e^{2\pi i jm/2^n} \quad .$$

The g_j, $1 \leq j \leq n$, are given by

$$g_j = 2^{n-j}, \quad 1 \leq j \leq n \quad .$$

This brings to mind the question of how a frequency shift keyed telemetry system with trigonometric FFT detection would compare with a Mariner '69 High-Rate type system. It turns out that the Mariner '69 FFT method yields a simpler machine because the characters of a direct sum of n groups of order 2 are all real; no trigonometric computations are involved. It appears in fact that the decoding procedure used in the Mariner '69 High-Rate Telemetry System is in some sense the simplest possible with a good code.

We shall now consider the FFT method adopted for Mariner '69. We must specify the group elements g_j, which in turn specifies the FFT. The character group $\hat{G}$ is now thought of as being isomorphic to the additive group of binary n-tuples

$$\{d = (d_n, d_{n-1}, \ldots, d_1)\}$$

and the group G itself to the same group:

$$G = \{g = (b_1, b_2, \ldots, b_n)\} .$$

Then a character $\chi^{(d)}$ operates on G by

$$\chi^{(d)}(b) = \phi^{-1}\left(\sum_{i=1}^{n} d_i b_i\right) ,$$

the sum being modulo 2 . The Payley-Walsh function w then corresponds to the n-tuple d whose $(n-i)^{\underline{th}}$ component is the exponent of the $i^{\underline{th}}$ Rademacher function (following the column indexing on the matrix K_n from 1 to n), in the representation of w as a product of Rademacher functions. Furthermore, lexicographic ordering of the characters $\chi^{(d)}$, by n-tuple d from the left, agress with the ordering of the rows of H_n from top to bottom. The element g_j is the n-tuple with all 0's, except for a 1 in position j . Thus,

$$S_0(\ell_1, \ell_2, \ldots, \ell_n) = f(g_1^{\ell_1} \cdots g_n^{\ell_n}) ,$$

$$S_1(\chi;\ell_2, \ldots, \ell_n) = S_0(0, \ell_2, \ldots, \ell_n) + \chi(g_1)S_0(1, \ell_2, \ldots, \ell_n) ;$$

for $1 \leq j \leq n-1$, define

$$S_{j+1}(\chi;\ell_{j+2}, \ldots, \ell_n) = S_j(\chi;0, \ell_{j+2}, \ldots, \ell_n) + \chi(g_{j+1})\, S_j(\chi;1, \ell_{j+2}, \ldots, \ell_n).$$

Thus,

$$\hat{f}(\chi) = S_n(\chi) .$$

Note that

$$\chi^{(d)}(g_j) = +1 \text{ or } -1$$

according as

$$d_j = 0 \text{ or } 1 .$$

Thus, $S_j(\chi;1, \ell_{j+1}, \ldots, \ell_n)$ is added to or subtracted from $S_j(\chi;0, \ell_{j+1}, \ldots, \ell_n)$ according as $d_j = 0$ or 1 . The notation $S_j(d;\ell_j, \ldots, \ell_n)$ will now be used instead of $S_j(\chi;\ell_j, \ldots, \ell_n)$.

This formulation of the FFT suggests the design of the "Green Machine" used to detect the Mariner '69 High-Rate Telemetry by the FFT technique. This machine not only evaluates the FFT, but is modular, because of the choice of the g_j . This means that the (32,6) Reed-Muller code can be expanded to the (64,7) Reed-Muller code by using the existing equipment and adding one stage of length 5 and one of length 6 . This expandability property is an important one for multi-mission telemetry, for it means that an entirely different machine need not be built for each slightly different mission.

In Figure VI, we see a schematic diagram of the Green Machine for $n = 3$; in general, $3 \cdot 2^n - 2$ units of storage are needed. A Green Machine for $n = 0$ is simply a single-stage register that accepts the input function f, defined only on 0, and stores it as f . The register then contains the Fourier Transform $\hat{f}$. For $n = 1$, the Green Machine consists of two single-stage registers and a two-stage register. The "upper" single stage register contains $f(1)$, the "lower" $f(0)$. The two-stage register is loaded from the two one-stage registers by putting $f(0) + f(1)$ in the second position, and $f(0) - f(1)$ in the first. The two-stage register then contains $\hat{f}(\chi^{(d)})$, the indexing being from right to left in increasing order on d . This ordering, is of course, the same ordering as the rows of the Hademard matrix H_1 .

We design and build an $(n+1)$-stage Green Machine as follows from an n-stage machine. We assume that we have an n-stage Green Machine which computes $\hat{f}(\chi^{(d)})$ in the correct order on d . Add to this machine one extra register of length 2^n, the so-called "upper register"; the original one of length 2^n is called the "lower register". Also add one register of length 2^{n+1} . The Green Machine of order $n+1$ works as follows. After $f(0), \ldots, f(2^n - 1)$ have been received, the Green Machine of order n computes $\hat{f}_0(\chi^{(d)})$ in the lower register in the proper order, where f_0 denotes f, originally defined on $\{0, 1, \ldots, 2^{n+1} - 1\}$, restricted to $\{0, 1, \ldots, 2^n - 1\}$; d is here an n-tuple. Then the register of lengths less than 2^n are set back to 0, and $f(2^n), f(2^n+1), \ldots, f(2^{n+1}-1)$ are received. The lower register of length 2^n is replaced by the upper register of length 2^n, and the Green Machine of order n now computes in proper order $\hat{f}_1(\chi^{(d)})$, where

$$f_1(m) = f(2^n + m), \quad 0 \leq m \leq 2^n - 1 .$$

After all 2^{n+1} values of f have been received, the register of length 2^{n+1} is loaded from the upper and lower registers as follows. The sum of the rightmost positions of the lower and upper register is loaded into the rightmost position of the register of length 2^{n+1}; the sum of the second rightmost positions into the second rightmost position, and so on, until the 2^n rightmost positions of the

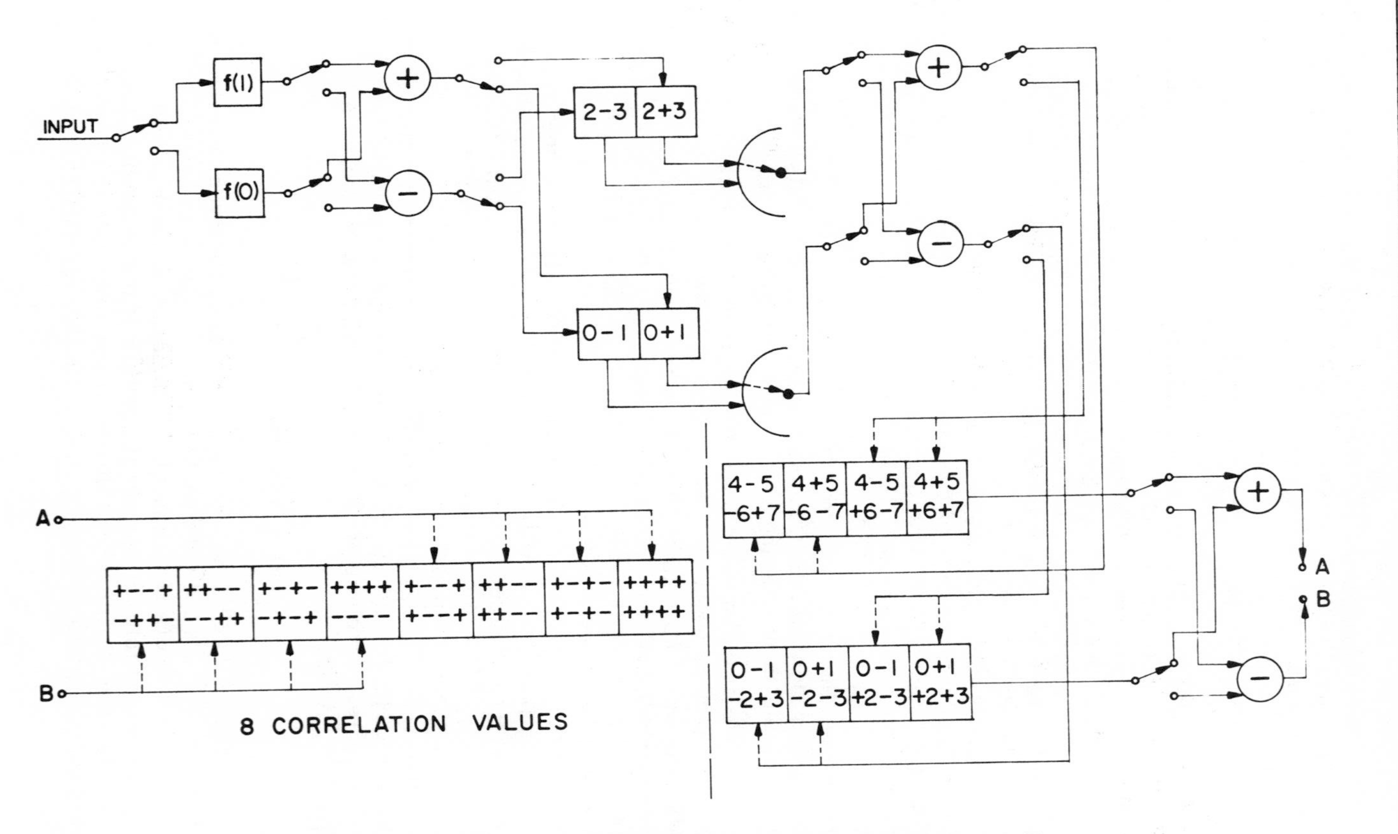

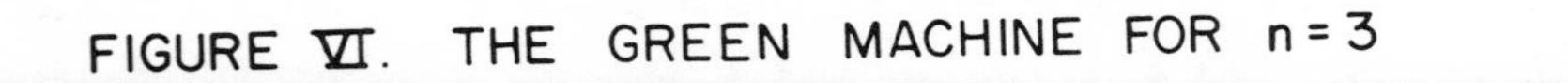

FIGURE VI. THE GREEN MACHINE FOR n = 3

register of length 2^{n+1} have been filled. The 2^n leftmost positions are now filled from the right with the differences of the bottom and top register, starting on the right. The process continues until the register of length 2^{n+1} has been entirely filled.

The length 2^{n+1} register does indeed now contain $\hat{f}(\chi^{(d)})$ in the correct order where the d here are now (n+1)-tuples with d_{n+1} placed on the left. For, the FFT rule is

$$S_{n+1}(d) = S_n(d;0) + \phi^{-1}(d_{n+1}) S_n(d;1) \quad .$$

Now the lower register contains $S_n(d,0)$ in proper order, and the upper $S_n(d;1)$. And $\phi^{-1}(d_{n+1})$ is +1 precisely when the leftmost (most significant) bit d_{n+1} of d is 0, that is, for the 2^n rightmost positions of the register of length 2^{n+1}. Thus, the Green Machine of order n+1 does indeed work as it should, and can be built in modular fashion.

The Green Machine as designed above could have been proven to perform the right correlations with the rows of H_n directly, without detouring through the FFT. But is exciting and instructive to witness an example where generalization pays off!

5. Word Synchronization

The Mariner IV telemetry system used bit-by-bit detection; hence, there was no problem of obtaining word synchronization before detection could take place. However, in the Mariner '69 High Rate System, word synchronization must be obtained before the most likely word can be outputted. Of course, subcarrier and then symbol sync have to be obtained first, but we begin the discussion assuming that sybol sync has already been achieved. After word sync has been obtained, frame sync can be done, but we are not concerned with that problem here either.

To obtain word synchronization, a coset of the (32,6) biorthogonal Reed-Muller code is used rather than the code itself. That is, a certain 32-tuple, called the comma-free vector, is added to the encoded 32-tuple on board the spacecraft, and, of course, removed on the ground so that FFT decoding can begin. During the word sync process, all 32 possible symbol positions are assumed in turn to be the correct start position. The Green Machine then goes into operation, after the comma-free vector is stripped off, and the correlation of largest absolute value is recorded. Also noted is a random other correlation. If these are not sufficiently different for several successive alleged words, then the next position is tried. The correct position is found with high probability when the correlation of largest absolute value is substantially larger than a random other correlation for several successive word times.

First let us review the definition of index of comma freedom. A code of length N is said to be comma-free of index r when, for every k, $1 \leq k \leq N-1$, every k-overlap, i.e. every N-tuple formed from the last k positions of one codeword followed by the first $N-k$ positions of another, possibly the same, codeword, differs from every codeword in at least r of the N positions. If $r \geq 1$, the code is called comma-free; however, we are interested in high indices of comma freedom.

Let us first give an upper bound on the index of comma freedom of any $(2^n, n+1)$ biorthogonal code, following Baumert and Rumsey [5]. The bound uses 3-overlaps. First note that in a biorthogonal code, given any three columns, all 8 3-tuples occur at those positions for some codeword. This result follows from the fact that by suitable row permutations, the $2^{n+1} \times 3$ matrix of the code restricted to the 3 positions can be made equal to the $2^{n+1} \times 3$ matrix of the leftmost 3 columns of the matrix $\phi^{-1}(K_{n+1})$.

Now let $N = 2^n$ be the code length, let $m = N-3$, and let y_m be an N-tuple consisting of 3 zeros followed by m entries of ± 1. Let $\{x_i, 1 \leq i \leq N\}$ denote an orthogonal subset of the $(2N, n)$ biorthogonal code. Let

$$\rho_i = (y_m, x_i)$$

denote the inner product, i.e., the correlation, between y_m and x_i. That is,

$$y_m = \sum_{k=1}^{N} \left(\frac{\rho_i}{N} \right) x_i \ ,$$

so

$$m = (y_m, y_m) = \frac{1}{N} \sum_{k=1}^{N} \rho_i^2 \ .$$

Hence, some ρ_i^2 exceeds m; since ρ_i is an odd integer, some ρ_i exceeds

$$2\{\frac{\sqrt{m}}{2} - 1\} + 1 \ ,$$

where $\{\ldots\}$ denotes the rounding upward function.

Now consider 3-overlaps v consisting of the last 3 positions of some codeword z with the first $N-3$ positions of some codeword y. Let y_m be the N-tuple whose first 3 positions are 0's , and whose last N-3 positions agree with the last N-3 positions of y . Then there is a codeword x (from among the x_i) such that the correlation ρ_1 between y_m and x satisfies

$$\rho_1 \geq 2\{\frac{\sqrt{m}}{2} - 1\} + 1 \ .$$

Let z be a codeword whose last 3 positions are the first 3 positions of x, and let v be the 3-overlap between z and y . Then the correlation

$$\rho = (v, x)$$

satisfies

$$\rho = \rho_1 + 3 \ ,$$

so that

$$\rho \geq 2\{\frac{\sqrt{m}}{2} - 1\} + 4 \ .$$

Now the 3-overlap v differs from the codeword x in

$$\frac{N - \rho}{2}$$

positions, hence the index of comma freedom of the $(2^{n+1}, n)$ biorthogonal code is upper bounded by

$$2^{n-1} - 2 - \{\frac{\sqrt{2^n - 3}}{2} - 1\} \ .$$

When n = 5, the upper bound is

$$16 - 2 - \{\frac{\sqrt{29}}{2} - 1\} = 14 - 3 = 11 \ .$$

Now the coset of the biorthogonal (32, 6) Reed-Muller code has index of comma freedom 6, and in fact there is even a coset of index 7 (but of no higher index). However, the possibility of FFT decoding is so important that no search has been made for biorthogonal (32, 6) codes of index of comma freedom greater than 7 . After all, a lower index of comma freedom merely means that more overlap words must be examined before one can be reasonably sure that one is not in sync.

The comma-free vector actually used in the Mariner '69 High-Rate Telemetry System is the 32 tuple

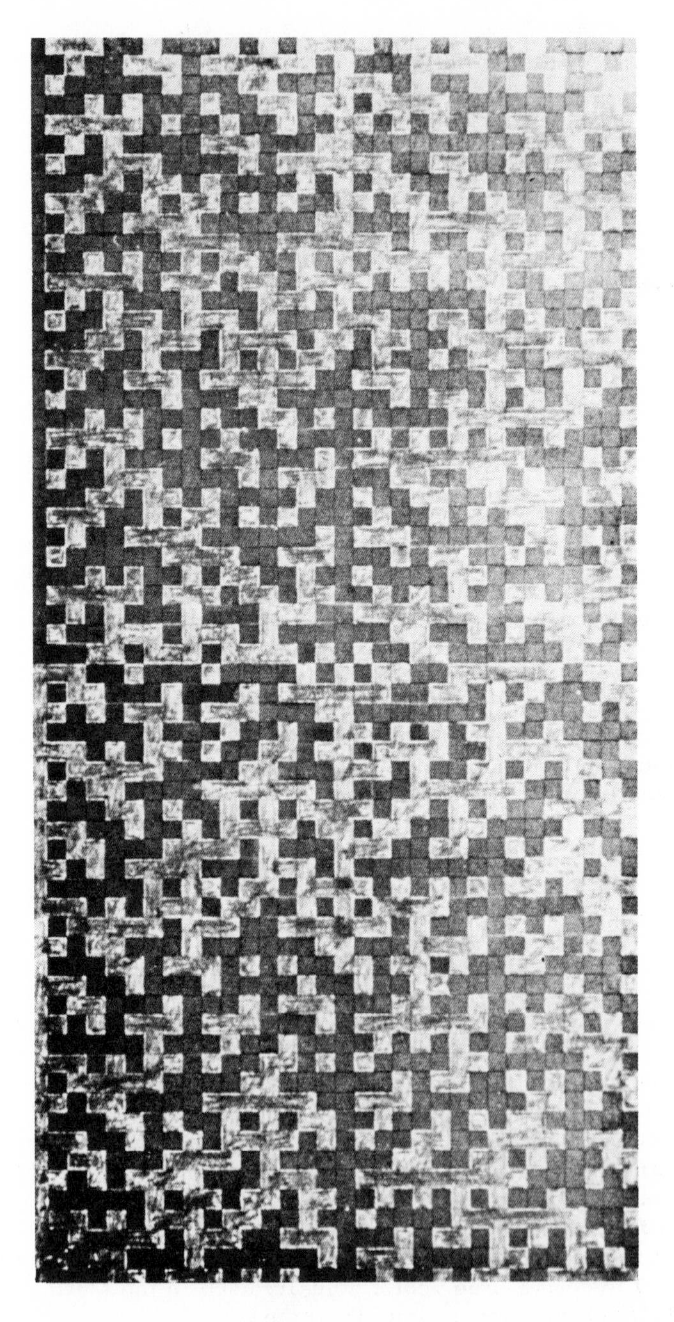

Fig. VII. The Code of Index 6

1000 1101 1101 0100 0010 0101 1001 1111 .

The resulting biorthogonal code is shown in Figure VII. This coset was shown to have comma-free index 6 by Stiffler in [6]. In [5], it was shown that no coset has index of comma freedom greater than 7, and one, in fact all, of index 7 were exhibited. Here is a comma-free vector of weight 8, the smallest weight possible for a comma-free vector where coset has index of comma freedom 7 :

1001 0011 0101 1000 0001 0000 0000 0000 .

The reason that Stiffler's coset of index 6 was chosen instead of one of index 7 is closely related to the way Stiffler found his comma-free vector. He took all (binary) maximal length shift register sequences of length 31, added one bit in all 32 possible positions, and tested the resulting coset by computer. The comma-free vector adopted consists of the maximal-length sequence whose initial condition is 00011, whose recursion polynomial is the primitive trinomial $x^5 + x^2 + 1$, and which is preceded by a 1 .

A simple device for performing the addition of this vector is shown in Figure VIII. The idea is that the 5 stage shift register keeps shifting until the five-tuple 11110 is detected by a word detector. When this happens, an extra 1 is interpolated into the output sequence before the leftmost 1 of the five-tuple 11110 outputs, and the shifting proceeds. This extra 1 is added to the leftmost position of the encoded word of the (32,6) Reed-Muller code, and after that the leftmost position of the register is added to the subsequent 31 positions at each shift of the register. When all 32 positions have been added in, the word has been placed into the proper coset, and is sent to the telemetry channel. At the same time, the register is again in its correct 11110 configuration for the next word. Thus, this procedure uses only 5 stages plus the word detector gate for the insertion of the extra 1; a simple-minded approach would use 32 stages to add the comma-free vector. Since equipment on the spacecraft and not encoding time was the limiting factor, this coset was adopted for use as the Mariner '69 High-Rate Telemetry code. None of the 2048 comma-free vectors yielding cosets of index 7 had any property that would simplify the process this much.

6. System Construction

For a good discussion of the digital hardware in the ground portion of the Mariner '69 High-Rate Telemetry System, see Winkelstein [7]. Much of the control is done by computer software, but the actual FFT and comma-free vector stripping is done by special-purpose

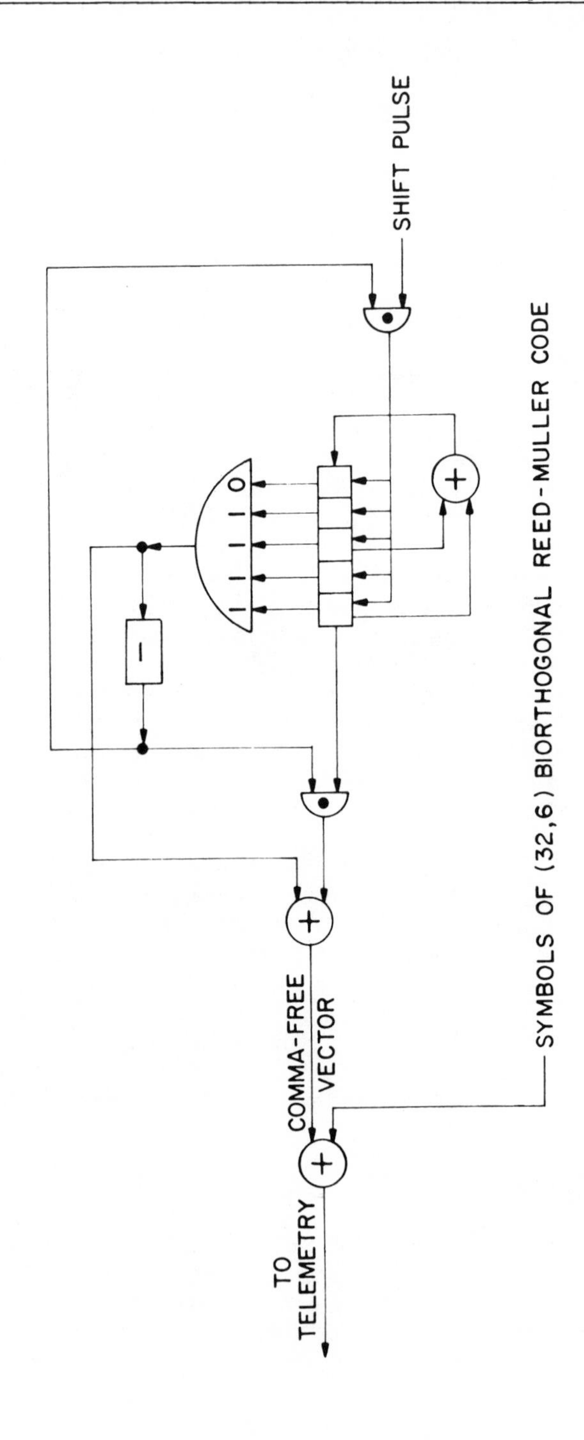

FIGURE VIII. ADDING THE COMMA - FREE VECTOR

Fig. IX. Lab Set A

digital hardware. Figure IX shows Lab Set A, a test version of the field equipment, built out of integrated circuits. Figure X shows some of the construction details: a card drawer with integrated circuits mounted on wide cards. The purpose of each card is indicated by a label on the cards. This laboratory set is being used for interface tests with other system hardware while the actual operational equipment is being built.

This operational equipment, Field Sets 1 and 2, is being built with the standard discrete-component logic cards used in the Deep Space Network. Figure XI shows a typical block of digital cards — this one contains the cards which strip the comma-free vector from a length 32 word so that FFT decoding can be done by the Green Machine portion of the equipment.

Tests conducted with Lab Set A show that the equipment functions as predicted by theory. When encounter of the Mariner '69 spacecraft with Mars takes place in August 1969, the Field Sets will begin the decoding of the actual digital television.

7. Areas for Further Research

This last section states some general problem areas in combinatorial telemetry suggested by the Mariner '69 High-Rate System. Some of the sought after solutions would help future versions of such a system; others of the problems are of interest in their own right.

I) Find other codes efficiently decodeable by a Fast Fourier Transform. Such decoding might even be useful on the binary symmetric channel.

II) Is the use of biorthogonal Reed-Muller codes with FFT decoding optimal in the sense of algorithm complexity?

III) Find a method for generating good comma-free vectors for biorthogonal Reed-Muller codes without exhaustive searching.

IV) Find a class of good comma free vectors which are easy to generate by simple finite-state machines.

V) Find a precise upper bound for the index of comma freedom for arbitrary cosets of biorthogonal Reed-Muller codes.

VI) Find a coset of the (64, 7) Reed-Muller code of largest index of comma freedom.

VII) Find a better upper bound for the index of comma freedom of any biorthogonal code.

VIII) Find a good lower bound on the index of comma freedom of the best coset of a biorthogonal Reed-Muller code.

We close this paper with the hope that, among other things, we have shown that combinatorial reasoning has participated in the

Fig. X. Integrated Circuit Card Drawer

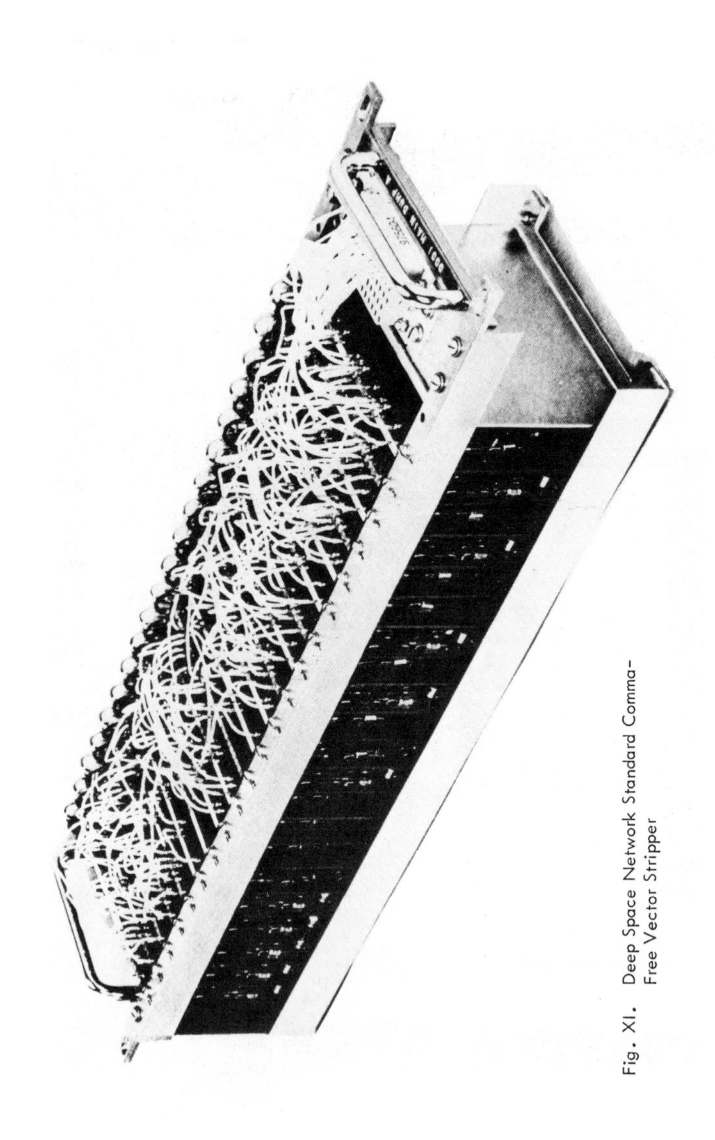

Fig. XI. Deep Space Network Standard Comma-Free Vector Stripper

invention and design of an important telemetry system from its very inception, and that, conversely, the existence of such a system has enriched combinatorial mathematics with new problems, techniques, and results.

List of Figures

References

[1] Viterbi, Andrew J., "Principles of Coherent Communication", McGraw Hill, New York, 1966.

[2] Posner, Edward C., "Properties of Error-Correcting Codes at Low Signal-to-Noise Ratios", SIAM Jour. Appl. Math., vol. 15 (1967), pp. 775-798.

[3] Welch, L. R., "Computation of Finite Fourier Series", Space Programs Summary No. 37-37, vol. IV (Feb. 1966), pp. 295-297, published by Jet Propulsion Laboratory, Pasadena, Calif.

[4] Green, R. R., "A Serial Orthogonal Decoder", ibid., No. 37-39, vol. IV (June 1966), pp. 247-251.

[5] Baumert, L. D., and H. C. Rumsey Jr., "The Index of Comma Freedom for the Mariner Mars 1969 High Data Rate Telemetry Code", ibid., No. 37-46, vol. IV (Aug. 1967), pp. 221-226.

[6] Stiffler, J. J., "Synchronization Techniques", Chap. Eight of "Digital Communications with Space Applications", S. W. Golomb, editor, Prentice-Hall, Englewood Cliffs, 1964.

[7] Winkelstein, R. A., "High Rate Telemetry Project Digital Equipment", Space Programs Summary 37-48, vol. II (Nov. 1967), pp. 107-114.

This paper presents the results of one phase of research carried out at the Jet Propulsion Laboratory, California Institute of Technology, under Contract No. NAS 7-100, sponsored by the National Aeronautics and Space Administration.

NEAL ZIERLER

Linear Recurring Sequences and Error-Correcting Codes

1. Introduction

The theory of linear recurring sequences over finite fields has many applications of both a theoretical and practical nature to error-correcting codes. Cyclic codes, that is, codes generated by linear recursions, have numerous interesting and useful properties; see, for example, references 1, 3, 4, 5, 6, 8, 10 and 11. The emphasis in the relevant parts of these works, with some important exceptions, is on the use of the theory of linear recurring sequences in constructing codes and exploring their weight distributions and other error-correcting characteristics. Our primary interest here is in the application to the decoding of error-correcting codes. This work was inspired to a large extent by E. R. Berlekamp's remarkable algorithm [1] for doing most of the computation in the method [5] due to D. Gorenstein and the writer for decoding the Generalized BCH codes. Another important source for us has been the paper of G. D. Forney [3] which showed that the Gorenstein-Zierler decoding method could be modified for the efficient correction of erasures as well as errors and also gave a simplified method for finding error values once the locations are known. These ideas suggested Lemma 13 of this paper and its application. They were also incorporated, along with Berlekamp's algorithm, into an experimental error-and-erasure-correcting coder-decoder developed by J. Terzian and the writer [10, 11]. I mention that work merely to suggest that certain of the ideas we are dealing with here can, in fact, be implemented in such a way as to provide error-free communication at information rates near channel capacity for certain real channels.

The theory of linear recurring sequences has been vigorously explored during the past decade [1, 2, 4, 5, 6, 7, 8, 9, 12]. Section 2 of this paper is a brief, almost self-contained, introduction to the theory in the form of a series of lemmas, together with some less familiar results, that we apply in Section 3. In Section 3.1 we consider the case in which each syndrome is a linear recurring sequence whose minimum polynomial specifies the error locations for a suitably restricted class of errors. Efficient procedures are described for finding the error locations and the error and erasure values. The Generalized BCH codes are obtained by a specialization of parameters. In Section 3.2 we consider briefly the opposite extreme; namely, the case in which every syndrome sequence has the same minimum polynomial.

2. Linear Recurring Sequences over Finite Fields

Let q be a prime power and let F be the field with q elements. Let S denote the set of all sequences $s = (s_0, s_1, \ldots)$ of elements of F and let $f(x) = c_n x^n + c_{n-1}x^{n-1} + \ldots + c_0$ be a polynomial with coefficients in F and $c_0 c_n \neq 0$. We say the sequence s is a linear recurring sequence generated by $f(x)$ if

$$c_n s_i + c_{n-1} s_{i+1} + \ldots + c_0 s_{i+n} = 0 \qquad \text{for } i = 0, 1, \ldots \quad (1)$$

We use $G(f)$ to denote the set of all linear recurring sequences generated by f. If K is an extension of F, we may also consider the set of linear recurring sequences of elements of K generated by f, since the coefficients of f belong to K, and denote the set of all such sequences by $G_K(f)$. Clearly $G(f) = G_F(f) \subset G_K(f)$.

A sequence s is periodic if for some positive number r, $s_i = s_{i+r}$ for $i = 0, 1, \ldots$. Every linear recurring sequence is periodic since the linear transformation $(s_i, s_{i+1}, \ldots, s_{i+n-1}) \rightarrow (s_{i+1}, s_{i+2}, \ldots, s_{i+n})$ of $V_n = X_n F$ to V_n defined by (1) is nonsingular. Conversely, every periodic sequence is in $G(f)$ for some f (with $f = x^r - 1$, if r is a period, for example), so the set of all such f is a nonempty ideal in $F[x]$. We have

Lemma 1. Let s be a periodic sequence. Then there exists a polynomial $f(x)$ such that $s \in G(g)$ if and only if $f | g$.

The polynomial f of the lemma is clearly unique and is called the minimum polynomial of s.

Let L be the shift mapping from S to S:

$$(Ls)_i = s_{i+1}, \qquad i = 0, 1, \ldots .$$

Lemma 2. Let R be a subset of S. $R = G(f)$ for some f if and only if

(i) The members of R are periodic.

(ii) R is a finite dimensional subspace of S over F.

(iii) R is closed under L.

Remark. Evidently the polynomial f of Lemma 2 is unique and dimension R = degree f.

It is convenient to modify the notation as follows. Regard the members of S as formal power series. Thus, the shift mapping now takes the form

$$(Ls)(x) = s_1 + s_2x + \ldots$$

and since power series may be multiplied by polynomials, we restate (1) as

$$L^n fs = 0 \ . \tag{2}$$

Identifying the periodic members of S with elements of the field $F(x)$ of rational functions we have

$$G(f) = \{g/f \mid \text{degree } g < \text{degree } f\} \ . \tag{3}$$

It is easy to see that f is the minimum polynomial of g/f if and only if $\gcd(f, g) = 1$. If $\{R_i\}$ are a finite number of subsets of S we let $\Sigma R_i = \{\Sigma s_i \mid s_i \in R_i\}$. Let $\{f_i\}$ be a finite family of polynomials. One proves easily:

Lemma 3.

(i) $\Sigma G(f_i) = G(\mathrm{lcm}\{f_i\})$

(ii) $\bigcap G(f_i) = G(\gcd\{f_i\})$.

Corollary 1. $f \mid g$ if and only if $G(f) \subset g(g)$.

Corollary 2. Let $f(x) = \Pi f_i^{r_i}(x)$ where f_i are distinct irreducible polynomials. Then $G(f) = \Sigma G(f_i^{r_i})$.

Lemma 3 has an important consequence. Suppose $f \in F[x]$ is irreducible of degree $n > 1$ and let $K = F(a)$, a a root of f. Since $f(x) = \Pi_{i=0}^{n-1}(x - a^{q^i})$, $G_K(f)$ splits into $\Sigma_{i=0}^{n-1} G_K(x - a^{q^i})$.

Let b be a nonzero element of K. The members of $G(x - b)$ are all power series

$$c + cb^{-1}x + cb^{-2}x^2 + \ldots$$

for $c \in K$ so a member s of $G_K(f)$ is specified by choosing n members $c_1, \ldots, c_n$ of K and then

$$s_j = \Sigma c_i (a^{-q^i})^j, \qquad j = 0, 1, \ldots \ . \tag{4}$$

Evidently, $s_j \in F$ for all j if and only if the c_i are conjugate over F. Hence, defining trace: $K \to F$ by

$$\text{trace } b = \sum_{i=0}^{n-1} b^{q^i}$$

we have

Lemma 4. Suppose f is irreducible in $F[x]$ and let a be one of its roots in an extension of F. An isomorphism $s \leftrightarrow c$ of the F-modules $G_F(f)$ and $F(a)$ is defined by

$$s_j = \text{trace}(ca^{-j}), \qquad j = 0, 1, \ldots \quad . \tag{5}$$

If s is a periodic member of S, let $p(s)$ denote its least period. For $f \in F[x]$, let $p(f)$, the period of f, be defined by

$$p(f) = \text{lcm}\{p(s) \mid s \in G(f)\} \quad .$$

It is easy to see that if f is of degree n, the sequence $s \in G(f)$ with $s_0 = s_1 = \ldots = s_{n-2} = 0$, $s_{n-1} = 1$ has f as minimum polynomial.

Lemma 5. Every polynomial f is the minimum polynomial of some sequence.

Lemma 6. Suppose f is the minimum polynomial of the sequence s. Then $p(s) = p(f)$.

Proof. Since $p(s)$ is a period of s, $s \in G(x^{p(s)} - 1)$. Then by Lemma 1, $f \mid x^{p(s)} - 1$, so $G(f) \subset G(x^{p(s)} - 1)$ by Lemma 3. Hence $p(t) \mid p(s)$ for all $t \in G(f)$ and $p(s) = p(f)$.

Corollary. $p(f) = \max\{p(s) \mid s \in G(f)\}$.

Lemma 7. If $s^i \in G(f_i)$ and the f_i are pairwise relatively prime, then $p(\Sigma s^i) = \text{lcm}\{p(s^i)\}$.

Proof. Let $r = p(\Sigma s^i)$. Clearly $r \mid \text{lcm}\{p(s_i)\}$. Now $\Sigma s^i_j = \Sigma s^i_{j+r}$ for all r, so $s^1_j - s^1_{j+r} = \Sigma_{i>1} s^i_{j+r} - s^i_j$ belongs to $G(f_1) \cap G(\Pi_{i>1} f_i)$. Since f_1 is prime to $\Pi_{i>1} f_i$, this intersection contains only the zero sequence by Lemma 3, part (ii). Hence r is a period of s^1; i.e., $p(s^1) \mid r$. Similarly $p(s^i) \mid r$ for all i, so $\text{lcm}\{p(s_i)\} \mid r$ and the equality follows.

By the order of a nonzero member of a finite field, we mean its order as a member of the multiplicative group of the field.

By Lemma 4, the period of an irreducible polynomial is the ord of its roots (see [12, Theorem 4] for details in the case of multiple roots).

Let us now fix an integer $r > 1$ and let S_r denote the set of all periodic members of S with period dividing r. Clearly $s \in S_r$

if and only if the minimum polynomial of s divides $x^r - 1$. We define the bilinear function $(,)$ from $S_r \times S_r$ to F by

$$(s, t) = \sum_{i=0}^{r-1} s_i t_i$$

and say $s \perp t$, s is orthogonal to t, if $(s, t) = 0$.

For $f(x) \in F[x]$, let $f^*(x)$ denote the polynomial whose roots are the inverses of the roots of f. If f is of degree n and $f(x) = \Sigma c_i x^i$, then

$$f^*(x) = x^n f(x^{-1}) = \Sigma c_i x^{n-i} .$$

Let $P_r = P_r(F)$ denote the set of all divisors of $x^r - 1$ over F. If $f \in P_r$, and g is any polynomial of degree less than $r - n$ then $g(x)f^*(x)(1 + x^r + x^{2r} + \ldots) = g(x)f^*(x)/(1 - x^r)$ is a member of S_r and from (1) $s \in G(f)$ implies

$$s \perp gf^*/(1 - x^r) . \qquad (6)$$

Conversely, if s is a periodic member of S for which (6) holds for all g of degree less than $n - r$, then $s \in G(f)$. We use the notation $T^\perp$ to denote the set of all members s of S_r such that $s \perp t$ for all elements t of the subset T of S_r. We have

Lemma 8. Let r be a positive integer and let f be a polynomial whose period divides r. Then in S_r,

$$G(f)^\perp = G((1 - x^r)/f^*) .$$

Corollary. Let K be an extension of F, r a positive integer and $f \in K[x]$ with period r (i.e., $f | x^r - 1$). Then the subset of $G_K(f)^\perp$ consisting of sequences with elements in F is

$$G_F((1 - x^r)/g^*)$$

where g is the polynomial of smallest degree in $F[x]$ that f divides.

Let us change the point of view a little now and suppose given $s \in S$ and the information that $s \in G(f)$ for some f of degree n, $f(x) = \Sigma_{i=0}^{n-1} c_i x^i$. If we wish to find the unknown coefficients c_1, $\ldots, c_n$ — having set $c_0 = 1$ for convenience — we have from (1) the equations

$$\sum_{i=1}^{n} c_i s_{j-i} = -s_j, \quad j = n, n+1, \ldots . \qquad (7)$$

Now the degree d of the minimum polynomial of s is $\leq n$, and the equations (7) are linearly independent for any d consecutive values of j, while any $d+1$ such equations are dependent, so that this property characterizes the degree of the minimum polynomial of a sequence.

Lemma 9. Suppose f is a polynomial of degree n, and is the minimum polynomial of the sequence s. Then the coefficients of f may be recovered from any $2n$ consecutive terms of s.

Berlekamp's algorithm [1] now takes the following form:

Lemma 10. Let $s(x) \in S$ have minimum polynomial $f(x)$ of degree n. Define two sequences of polynomials $\sigma_j(x)$ and $\tau_j(x)$ and a sequence of integers η_j as follows: $\sigma_0(x) = 1$, $\tau_0(x) = x$ and $\eta_0 = 0$. For $j = 0, 1, \ldots$, let a_{j+1} be the coefficient of x^j in $\sigma_j(x)s(x)$. Set

$$\sigma_{j+1} = \sigma_j - a_{j+1}\tau_j$$

$$\tau_{j+1} = \begin{cases} x a_{j+1}^{-1}\sigma_j & \text{if } a_{j+1} \neq 0 \text{ and } \eta_j \geq 0 \\ x\tau_j & \text{if } a_{j+1} = 0 \text{ or } \eta_j < 0 \end{cases}$$

$$\eta_{j+1} = \begin{cases} -\eta_j & \text{if } a_{j+1} \neq 0 \text{ and } \eta_j \geq 0 \\ \eta_j + 1 & \text{if } a_{j+1} = 0 \text{ or } \eta_j < 0 \end{cases} .$$

Then $\sigma_{2n-1}(x) = f(x)$.

Remark. At the completion of the jth step, the algorithm has used the terms $s_0, \ldots, s_{j-1}$ of s, so it does in fact deduce the minimum polynomial of degree n from $2n$ consecutive terms of the sequence. If the process is continued past step $2n-1$, all $a_{j+1} = 0$ for $j = 2n-1, 2n, \ldots$. Hence if ℓ consecutive terms of any sequence s are given, the minimum polynomial of s can be found providing its degree is at most $[\ell/2]$.

Lemma 11. Suppose f and g of degrees m and n respectively are relatively prime and s is a sequence of the form $t + u$ where t has f as minimum polynomial and $u \in G(g)$. Then f is the minimum polynomial of $L^n gs$.

Proof. $L^n gs = L^n g(t+u) = L^n gt + L^n gu = L^n gt$ since $L^n gu =$ by (2). Now $L^n gt \in G(f)$ by Lemma 2, parts (i) and (ii), so its minimum polynomial $\hat{f}$ divides f. Choose the polynomial g_2 of degree less than n so that $g_2 + gt \in G(\hat{f})$ and $(g_2 + gt)_{j+n} = (L^n gt$ $j = 0, 1, \ldots$. According to (3) we may write $t = f_1/f$ with $\gcd(f_1,$

and $g_2 + gt = h_1/\hat{f}$. Then $h_1/\hat{f} = g_2 + gt = g_2 + gf_1/f$ so $fh_1 = f\hat{f}g_2 + \hat{f}gf_1$. Hence $f \mid \hat{f}gf_1$. But f is prime to both g and f_1, so $f \mid \hat{f}$, and f is the minimum polynomial of $L^n gs$.

Lemma 12. Let f and g be relatively prime polynomials of respective degrees m and n . Suppose that g (and n) are known, that f (and m) are unknown, and we are given ℓ terms $s_0, \ldots, s_{\ell-1}$ of a sequence s which is the sum of a sequence with f as minimum polynomial and a sequence belonging to $G(g)$. Then f can be found providing $\ell \geq n + 2m$.

Proof. $L^n gs$ gives us $\ell - n$ consecutive terms of a sequence which has f as its minimum polynomial by Lemma 11. Hence we may apply Lemma 10 to find f providing $\ell - n \geq 2m$.

Lemma 13. Suppose f and g are known relatively prime polynomials of positive degrees m and n respectively and $\ell \geq m + n$. Suppose further that we are given terms $u_0, \ldots, u_{\ell-1}$ of a sequence u with f as minimum polynomial and ℓ terms of a sequence s of the form $au + v$ where a is an unknown member of F and v is an unknown member of $G(g)$. Then a may be found as follows. Let $t = L^n gu$ and let $w = L^n gs$. If $a \neq 0$, then for some $j = 0, 1, \ldots, m-1$, $t_j w_j \neq 0$, and for every such j, $a = w_j t_j^{-1}$. If $a = 0$, then $0 = w_0 = \ldots = w_{m-1}$.

Proof. First observe that since $\ell \geq m + n$, we have at least m terms of t and w . Now

$$
\begin{aligned}
w = L^n gs &= L^n g(au + v) = aL^n gu + L^n gv \\
&= aL^n gu \quad \text{since } v \in G(g) \\
&= at \ .
\end{aligned}
$$

Hence all $w_i = 0$ if $a = 0$. Suppose $a \neq 0$. Then, by Lemma 11, w has f as minimum polynomial, so at most $m - 1$ consecutive terms of w are zero. Since $w_j = at_j$ for all j and $a \neq 0$, $w_j \neq 0$ implies $t_j \neq 0$ and $a = w_j t_j^{-1}$.

Corollary. Suppose $r > 1$ and $f_1, \ldots, f_r$ are pairwise relatively prime polynomials of respective positive degrees m_i . Let $\ell \geq \Sigma_{j=1}^r m_j$ terms of a sequence u^i with f_i as minimum polynomial and unknown members a_i of F be given for each $i = 1, \ldots, r$. Suppose further that we are given ℓ terms of the sequence

$$
s = \sum_{i=1}^{r} a_i u^i \ .
$$

Then a_1 (and similarly any a_i) may be found as follows. Let

$g = \Pi_{i>1} f_i$, $n = \Sigma_{i>1} m_i$, $t = L^n gu^1$ and $w = L^n gs$. If $a_1 \neq 0$, $t_j w_j \neq 0$ for some $j = 0, \ldots, m_1 - 1$ and for any such j, $a_1 = w_j t_j^{-1}$. If $a_1 = 0$, $0 = w_0 = \ldots = w_{m-1}$.

An alternative method for finding all the unknowns a_i is as follows. Having found a_1 as above, we have $\ell_1 = \ell - m_1$ terms of the sequences

$$v^i = L^{m_1} f_1 u^i, \qquad i = 2, \ldots, r$$

and of $s^1 = L^{m_1} f_i s$.

Now $s^1 = \Sigma_{i=2}^r a_i v^i$, and we let $n = \Sigma_{i>2} m_i$, $g = \Pi_{i>2} f_i$, $t = L^n gu^2$ and $w = L^n gs^1$. Then if $a_2 \neq 0$, $t_j w_j \neq 0$ for some $j = 0, \ldots, m_2 - 1$, and for all such j, $a_2 = w_j t_j^{-1}$. Iteration of this procedure then yields all the a_i.

3. Applications to Error-Correcting Codes

We begin with the following notation and definitions.

- q a prime power
- F the field with q elements
- n a positive integer
- $V = V_n = V_n(F)$ the Cartesian product of n factors F
- k a positive integer, $k \leq n$
- A a k-dimensional subspace of V
- $(u, v) = \Sigma u_i v_i$ for u and v in V
- $W \subset V$
- $[W]$ = smallest subspace of V containing W
- $W^\perp = \{v \in V | (v, w) = 0 \text{ for all } w \in W\}$.

Clearly $W^\perp$ is a subspace of V of dimension $n - \dim[W]$ and $W^{\perp\perp} = [W]$. In particular, $A^\perp$ is an n-k-dimensional subspace of V and A is determined as the set of vectors u such that $(u, v^i) = 0$ for vectors $v^1, \ldots, v^\ell$ spanning $A^\perp$.

If a member u of A is transmitted and $v = u + e$ is received, the sequence $s_i = (v, v^i) = ((u+e), v^i) = (u, v^i) + (e, v^i) = (e, v^i)$, $i = 1, \ldots, \ell$ is called the syndrome of the error e. If one wishes to have $A = G(f)$ for some f, Lemma 8 tells us how we may choose the v^i if f is given or what f is if the v^i are given. This application of the theory is fully described elsewhere (see [1, 5]).

3.1. Consider now the class of codes with the property that every syndrome is a partial linear recurring sequence whose generating polynomial specifies the error locations whenever the error-and-erasure-vector belongs to a suitably restricted subset of V . This situation may be described in greater detail as follows: Let $\{f_i\}_{i=1}^n$ be pairwise relatively prime members of $F[x]$, let ℓ be a positive integer, and for each $i = 1, \ldots, n$ let $u_1^i, \ldots, u_\ell^i$ be ℓ terms of a sequence u^i with f_i as minimum polynomial. We define ℓ members $v^1, \ldots, v^\ell$ of V by $v_i^j = u_j^i$. Let $A = \{v^1, \ldots, v^\ell\}^\perp$. Suppose $a \in A$ is transmitted. When a is received and demodulated it appears in the form of a list w of n objects each of which either belongs to F or is an indeterminate. If w_i is an indeterminate, an erasure has occurred at position i . Let E be the set of positions $i = 1, \ldots, n$ at which erasures have occurred. Since these are known, so is the polynomial $g = \Pi_{i \in E} f_i$ of degree $\nu = \Sigma m_i$ where m_i = degree f_i . Let $\bar{w}$ be the element of V obtained from w by replacing each of its indeterminates by some member of F which, for convenience, we take to be 0 . Let $s = (s_1, \ldots, s_\ell)$ be the syndrome of $\bar{w}$, i.e., $s_j = \Sigma \bar{w}_i v_i^j$. We wish to find the error vector $e = \bar{w} - a$. Let D denote the set of positions i not in E for which $e_i \neq 0$. Our first objective is to find D . Now

$$s_j = \sum_{i=1}^{n} e_i v_i^j = \sum_{i=1}^{n} e_i u_j^i$$

$$= \sum_{i \in D} e_i u_j^i + \sum_{i \in E} e_i u_j^i \; .$$

Since $e_i \neq 0$ for $i \in D$, s is the sum of a sequence with $f = \Pi_{i \in D} f_i$ as minimum polynomial and a sequence belonging to $G(g)$. Let $m = \Sigma_{i \in D} m_i$. This is precisely the situation described in Lemma 12, so we can find f by the method described there providing $\ell \geq \nu + 2m$. That is, we compute $L^\nu ga$, giving us $\ell - \nu$ terms of a sequence which has f as its minimum polynomial, so that the algorithm of Lemma 10 may be applied to find f . To find D we must now determine which f_i divide f . An efficient means of doing this is to reduce f modulo each f_i with $i \notin E$ since $f \equiv 0 \pmod{f_i}$ if and only if $f_i | f$. Depending on the precise situation, it may be worthwhile to invest the extra time and/or hardware to find f/f_i instead, since the amount of computation remaining to be done is thereby reduced each time a divisor is found.

We have assumed that $\ell \geq \nu + 2m$ in order to find D, so, a fortiori, the condition $\ell \geq \nu + m$ is satisfied and we may therefore find the unknown error values e_i, $i \in E \cup D$, by either of the two procedures described in the corollary to Lemma 13. Specifically, let r be the number of elements of $E \cup D = \{i_1, \ldots, i_r\}$, let $h = f_{i_2} \cdots f_{i_r}$, $\mu = m_{i_2} + \ldots + m_{i_r}$, $t = L^{\mu} hu^{i_1}$ and $z = L^{\mu} gs$. If $e_{i_1} = 0$ (which may occur if $i_1 \in E$), then $0 = z_1 = \ldots = z_{m_{i_1}}$. If $e_{i_1} \neq 0$, then $t_j z_j \neq 0$ for some $j = 1, \ldots, m_{i_1}$ and for any such j, $e_{i_1} = z_j t_j^{-1}$,

If all $m_i = 1$, a number of simplifications result. In particular, μ and m are just the number of erasures and errors. A special case is

Example 1. The Generalized BCH Codes

Notation and definitions (continued).

n prime to q
b a primitive n^{th} root of unity in an extension of F
$K = F(b)$
$f_i = x - b^{-i}$, $i = 1, \ldots, n$
u^i = a nonzero member of $G_K(f_i)$, $i = 1, \ldots, n$.

If μ erasures and m errors occur, the resulting syndrome sequence is part of a linear recurrence generated by a polynomial of degree $\mu + m$; namely, the product of the appropriate f_i. Hence the error locations and the values of the errors and erasures may be found providing $\ell \geq \mu + 2m$. It is customary for these codes to also require each "vertical sequence" $\{v_i^j\}_{i=1}^n$ be generated by first degree polynomials. The most general solution to the two conditions is as follows: there exist nonzero b_0 and c_0 in K such that

$$f_i = x - (b_0 b^i)^{-1}$$

$$u_1^i = c_0 b^i .$$

3.2. In the preceding section, the polynomials f_i were assumed pairwise relatively prime. Here we make a few remarks on the opposite extreme. Let f be an irreducible polynomial and let $f_i = f$ for $i = 1, \ldots, n$. Let b be a root of f in an extension of F and let $K = F(b)$. According to Lemma 4, for each of the sequences $u^i \in G(f)$, there is a member c_i of K such that $u_j^i = \text{trace } c_i b^{-j}$ for $j = 1, 2, \ldots$. Consider the syndrome $s = s_1, \ldots, s_\ell$ that results from an error vector $y_1, \ldots, y_n$.

$$s_j = \sum_{i=1}^{n} y_i u_j^i = \Sigma y_i \text{ trace } c_i b^{-j} .$$

Since trace is a morphism of F-modules: $K \to F$,

$$s_j = \text{trace}\, \Sigma y_i c_i b^{-j}, \quad j = 1, \ldots, \ell \ . \tag{8}$$

Let λ be the morphism of F-modules: $V_n(F) \to K$ defined by

$$\lambda(y) = \sum_{i=1}^{n} y_i c_i \tag{9}$$

and let $\gamma: K \to V_\ell$ be the morphism of F-modules

$$\gamma(\beta) = (\text{trace}\, \beta b^{-1}, \ldots, \text{trace}\, \beta b^{-\ell}) \ .$$

We have the diagram

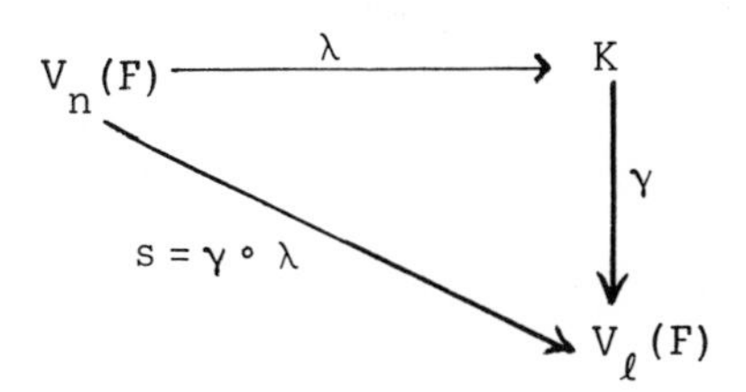

The code A is the kernel of s, which contains the kernel of λ . Indeed, it is not difficult to choose b so that γ is a monomorphism, and then $A = \text{kernel}\ \lambda$. A familiar requirement is that every nonzero member of A have at least d nonzero components, d a given integer, for then if any $(d-1)/2$ or fewer of the components of the received vector are in error, correction is possible. According to (9), this amounts to requiring that any set of d of the elements $c_1, \ldots, c_n$ of K be linearly independent over F .

Now suppose that we have fixed an element β of K and take $c_i = \beta^{i-1}$, $i = 1, \ldots, n$. Then (9) becomes

$$\lambda(y) = \sum_{i=1}^{n} y_i \beta^{i-1} , \tag{10}$$

and the kernel of λ is identified with the set of polynomials in F[x] of degree less than n divisible by the minimum polynomial of β in F[x] . Here we have a situation quite different from that of the preceding section, for now all the syndrome elements lie in F . For example, if $F = GF(2)$, the syndromes are, like the code vectors themselves, lists of elements of $GF(2)$. A familiar example is:

Example 2. The Golay Code. $f(x) = x^{11} + x^9 + x^7 + x^6 + x^5 + x + 1$ is irreducible over $GF(2)$ (of period 23), $n = 23$, $\ell = 11$, $\beta = b =$ a root of f, $u_j^i = \text{trace } b^{i-j-1}$. Since trace b^r is the coefficient of x^{10} in the minimum polynomial of b^r, trace $b^r = 0$ if b^r is conjugate to b, $= 1$ otherwise; i.e., $= 0$ if

$$r \in \{1, 2, 4, 8, 16, 9, 18, 13, 3, 6, 12\} ,$$

$= 1$ otherwise. Since any 11 successive powers of b are linearly independent over $GF(2)$, the matrix (u_j^i), $i = 1, \ldots, 23$; $j = 1, \ldots, 11$ has rank 11 and the dimension of A is $23 - 11 = 12$. The assertion that A is 3-error-correcting is equivalent to the assertion that every multiple of f of degree less than 23 shares with f the property of having at least 7 terms.

REFERENCES

[1] E. R. Berlekamp, "Algebraic coding theory," McGraw-Hill Book Co. Inc., New York, 1968.

[2] J. P. Fillmore and M. L. Marx, "Linear recursive sequences," to appear.

[3] G. David Forney, "On decoding BCH codes," IEEE Trans. Inf. Theory, 11(1965), 549-557.

[4] S. W. Golomb, "Shift register sequences," Holden-Day Inc., San Francisco, 1967.

[5] D. Gorenstein and N. Zierler, "A class of error-correcting codes in p^m symbols," J. Soc. Indust. Appl. Math., 9 (1961), 207-214.

[6] T. Kasami, S. Lin and W. W. Peterson, "Some results on cyclic codes which are invariant under the affine group and their applications," Information and Control 11(1967), 475-496.

[7] Dan Laksov, "Linear recurring sequences over finite fields," Math. Scand. 16(1965), 181-196.

[8] W. W. Peterson, "Error-correcting codes," M.I.T. Press, Cambridge, Massachusetts, 1961.

[9] Ernst S. Selmer, "Linear recurrence relations over finite fields," Dept. of Math., University of Bergen, Norway, 1966.

[10] J. Terzian and N. Zierler, "Evaluation of the MITRE 950 error-correcting coder-decoder," Digest of the 1966 IEEE Intern. Comm. Conf., June 1966, 30.

[11] J. Terzian, "Field test of the MITRE 950 coder-decoder," MITRE Tech. Rep. MTR-134, The MITRE Corp., Bedford, Massachusetts, February, 1966.

[12] N. Zierler, "Linear recurring sequences," J. Soc. Indust. Appl. Math., 7(1959), 31-48.

Institute for Defense Analyses
Princeton, New Jersey

E. R. BERLEKAMP

Block Coding for the Binary Symmetric Channel with Noiseless, Delayless Feedback

ABSTRACT

This paper considers the problem of transmitting messages reliably across a noisy binary symmetric channel which is accompanied by a noiseless, delayless feedback channel. Upper bounds are derived for the error correction capability of arbitrary block coding strategies, and explicit strategies are constructed which achieve these bounds at many rates. At certain rates, these strategies are asymptotically close-packed.

1. Background and Summary of Results

Led by Schalkwijk and Kailath (1966), communication theorists have recently displayed increasing interest in the problem of transmitting information across a noisy channel which is accompanied by a noiseless, delayless feedback channel of large capacity. Although Shannon (1956) showed that the capacity of the channel with feedback cannot exceed the capacity of the same one-way channel, the feedback channel nevertheless proves useful at all rates below capacity. Phenomenal reductions in the probability of error can be obtained when feedback is used with continuous channels which have an average power constraint at the transmitter, and substantial reductions can be obtained when feedback is used with discrete channels or continuous channels with peak-power constraints. Some of the feedback coding schemes which have been suggested for discrete channels [Horstein (1963)] employ a variable length stopping rule, according to which the transmitter continues sending digits until the receiver is "sufficiently sure" which message is intended. Although such schemes attain a low probability of error, they may occasionally require a relatively large number of digits to be transmitted before the receiver is able to reach a decision. For this reason, Shannon and Gallager suggested that it might be interesting to obtain error bounds for the best feedback coding strategy in which the number of digits to be transmitted is fixed in advance, i.e., the best block coding strategy. Several such bounds were obtained by Berlekamp (1964). The present paper extends some of those results which apply to the binary symmetric channel.

The present paper is devoted to the problem of determining n, the block length of the shortest feedback coding strategy which always enables the receiver to decide correctly among $M = 2^{Rn}$ possible transmitted messages, assuming that the channel alters no more than $e = fn$ of the transmitted bits. For very large n, the relationship between the rate R and the error-correction fraction f is shown in Figures 1 and 2.

The asymptotic relationship between f and R for the best one-way binary codes lies somewhere within the shaded region of Figure 1. The best known asymptotic lower bound on f as a function of R is obtained by a nonconstructive argument due to Gilbert (1952); the best upper bound is obtained by an argument due to Elias. Figure 1 also includes a weaker upper bound due to Hamming (1950). Proofs and generalizations of all of these bounds for one-way codes may be found in Chapter 13 of Berlekamp (1968).

The asymptotic relationship between f and R for the best binary feedback block coding strategies is shown in Figure 2. These upper bounds on $f(R)$ are derived in Part 3 of this paper; these lower bounds, in Part 4. At rates in the interval $.2965 < R < 1$, the best asymptotic upper bound on f is the volume bound, which is a generalization of Hamming's bound for one-way codes. For lower rates, $0 < R < .2965$, the best upper bound on $f(R)$ is a straight line which emanates from the point $R = 0$, $f = 1/3$, and joins the volume bound tangentially at $R = .2965$. One of the constructions presented in Part 4 of this paper yields a range of coding strategies which attain the rate and error correction fraction of any point on this straight line. Other constructions yield ranges of coding strategies which attain the rates and error-correction fractions of any point on the straight lines emanating from the point $R = 0$, $f = 1/t$, where t is any integer. These constructions yield asymptotically "close-packed" codes at any of the rates at which one of these straight lines touches the volume bound. Many compelling heuristic arguments suggest that it is actuall possible to obtain asymptotically close-packed feedback coding strategies at all rates $\geq .2965$. We give a construction which apparently yields codes along the straight line emanating from $R = 0$, $f = 2/7$ and joining the volume bound tangentially.

A comparison of Figures 1 and 2 reveals that the error-correction capability of the best binary feedback block coding strategies is asymptotically superior to the error-correction capability of the best one-way binary block codes, at all rates $0 \leq R < 1$. It is also worth noting that the lower bounds of Figure 2 are obtained by evaluating the

Figure 1. Asymptotic error-correction fraction of one-way BSC

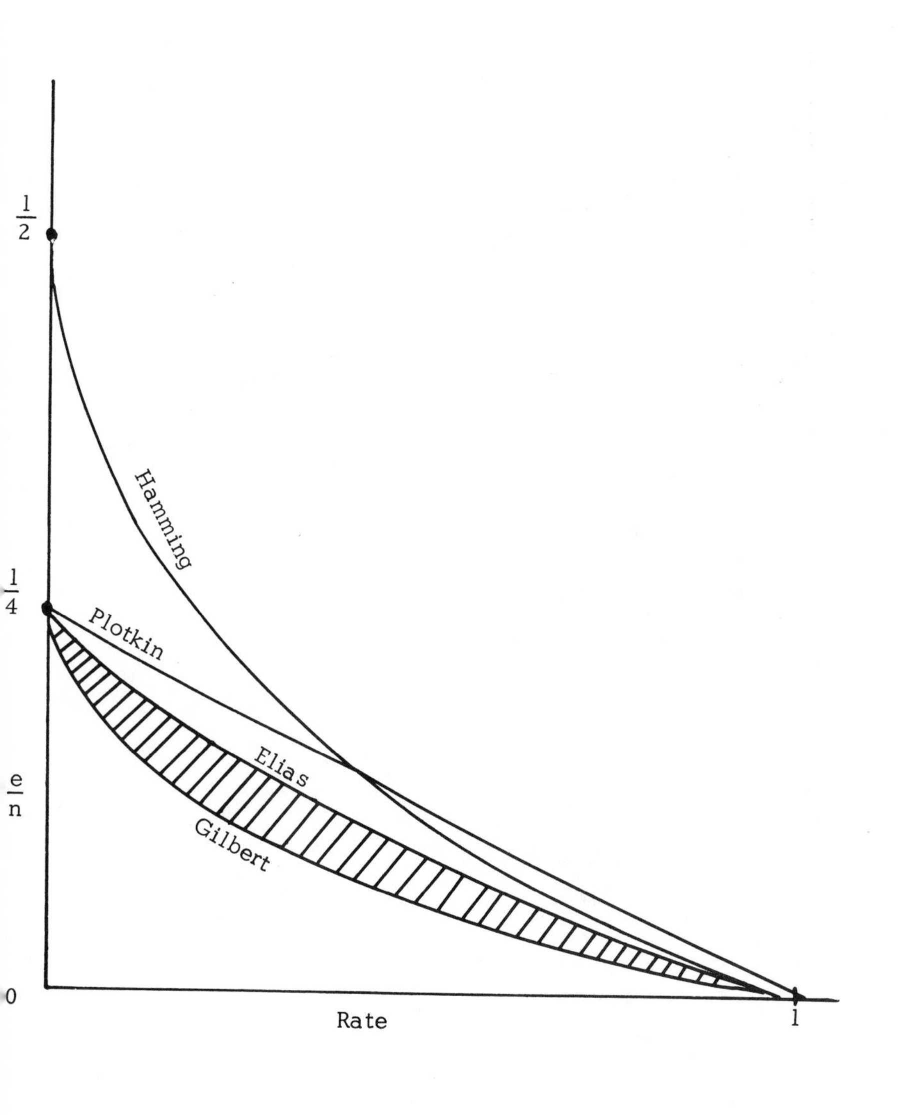

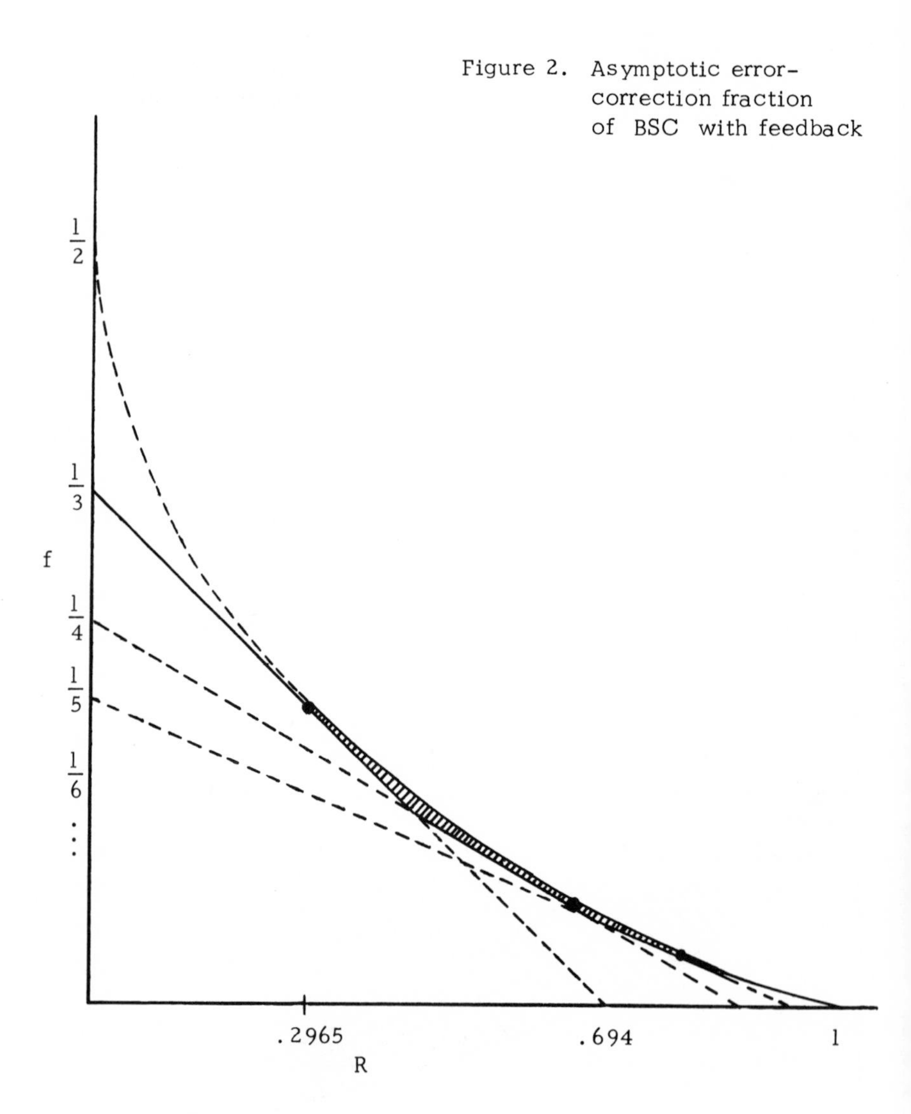

Figure 2. Asymptotic error-correction fraction of BSC with feedback

performance of certain explicitly stated strategies, whereas no known "explicit" constructions for one-way codes attain an asymptotically positive f at any positive R.

Before presenting the asymptotic upper and lower bounds on $f(R)$ in Parts 3 and 4 respectively, we introduce some preliminary notions in Part 2.

2. Coder vs Nature; States and Partitions

We now consider the problem of coding for the binary symmetric channel (Figure 3) with noiseless, delayless feedback.

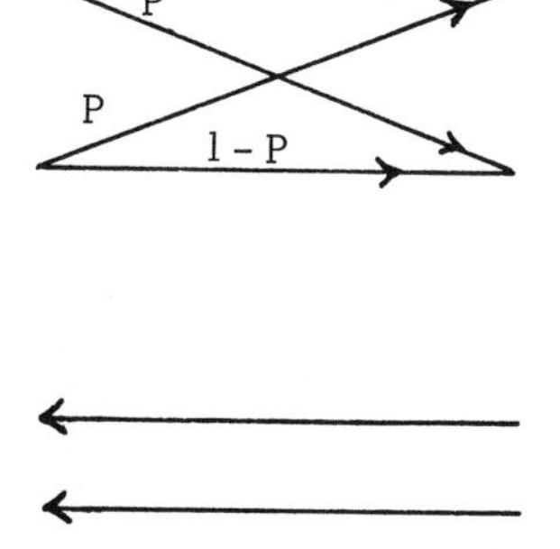

Figure 3. Binary symmetric channel with noiseless delayless feedback

We will often refer to the coding process as a game between two hostile opponents: Coder, a partnership including the transmitter and the receiver, and Nature, who controls the channel transitions.

The game of transmitting one block of information on this channel is played as follows: The source first selects one message from an ensemble of M equiprobable words. He attempts to convey this choice to the receiver by transmitting n bits across the noisy channel. Some of these transmissions may depend on information the source received from the feedback channel as well as on the selected message word.

We adopt the point of view that just prior to each forward transmission the receiver asks the source a yes-no question: "Is the correct message among the set S_i?" (S_i is a subset of the M possible messages.) The question is transmitted back to the source over the noiseless feedback link, and the source's answer is then sent to the receiver via the noisy channel. The source receives a noisy answer, and then asks another question. At each stage the questioned set, S_i, may depend on the entire past history of the game.

It may first appear that this viewpoint necessitates an unusually large amount of feedback, since at each stage the receiver transmits back a subset S_i, which may be any subset of M possibl message words. This transmission seems to require $\log_2 M$ bits of noiseless feedback. Actually, however, only one bit of feedback is required for each bit transmitted, because the transmitter may be endowed with the same deterministic subset-selecting machine as the receiver. The only inputs into this subset-selector are the results of previous questions, i.e., the received sequence of bits. Thus th evolution of the questioning process is determined only by the receiv sequence of answers. If the feedback channel can accommodate one noiseless bit for each noisy bit sent down the forward channel, the source can be kept informed of the received sequence. He then know as much as the receiver, and additional feedback cannot be of any additional help.

Any strategy may be viewed as a quiet-question, noisy-answ process of the type just described. One need only consider the set o possible selected messages which would cause the source to transmi a one next, and call this the questioned set. The question-answer viewpoint involves no restriction on the types of strategies permissible.

The receiver may regard each answer he receives as a vote against a certain subset of words. As the process proceeds, differen words acquire different numbers of unfavorable votes. After all n transmitted bits are received, the receiver must decide which word was transmitted. He obviously does best to select that word which has received the fewest unfavorable votes.

As an example, let us suppose that M = 8, n = 11. We denot the 8 possible messages by A, B, C, D, E, F, G, and H . We start with all 8 codewords having no votes against them, and 11 question remaining. The game might proceed as shown in Figure 4.

For example, when there were 5 questions remaining, the receiver asked the question: "Is the selected message among the set FEGH?" The reply that was received was "No".

In this game, Nature caused at least three channel errors. If it caused only three errors, than A was the selected message, and Nature's errors occurred at questions 10, 9, and 4. It is also possib that D was the selected message, in which case Nature caused four errors, at questions 7, 6, 3, and 1; or that F is the message, in whic case the four errors occurred at questions 11, 5, 2, and 1. If any of the other codewords was the message, then Nature committed six or more errors.

We now turn our attention to the problem of selecting a questioning strategy which guarantees that after all n questions have been asked, the receiver will be able to deduce correctly which word was transmitted unless the channel had made more than e errors. As

Votes Against	Questions Remaining: 11	10	9	8	7	6	5	4	3	2	1	0
0	ABCDEFGH	ADEH	DE	D	D							
1		BCFG	ACFH	CEF	F	DF	F					
2			BG	AGH	ACE	A	AD	ADF	DF	F		
3				B	BGH	BCEG	EG	-	A	AD	ADF	A
4						H	BC	BCEG	EG	-	-	DF
5							H	-	BC	BCEG	EG	-
6								H	-		BC	EGBC
7									-		-	-
8									H		-	-
9											H	H

Figure 4. A Sample Game

before, after each question we tally the number of negative votes against each possible message word. However, we may now throw away words which accumulate more than e unfavorable votes.

After each question, we record the number of words which have 0 negative votes, the number which have 1 negative vote, ... the number which have e negative votes. We write these numbers as components of a column vector, and call this vector the state of the game. If there are n questions remaining, this vector is called an n-state. The topmost components of this vector are often zeros. For this reason, we index the components from the bottom up and omit any zeros above the highest nonzero component:

$$\begin{matrix} \cdots \\ c_4 \\ c_3 \\ c_2 \\ c_1 \\ c_0 \end{matrix} = \underline{c}$$

The component c_i denotes the number of words which have received $e-i$ negative votes.

With $e = 3$, the states which occurred in the game of Figure 4 are shown in Figure 5.

Unused Votes Against	Questions Remaining											
	11	10	9	8	7	6	5	4	3	2	1	0
3	8	4	2	1	1	0	0	0	0	0	0	0
2		4	4	3	1	2	1	0	0	0	0	0
1			2	3	3	1	2	3	2	1	0	0
0				1	3	4	2	0	1	2	3	1

Figure 5. States of the game of Figure 4

At each question, the receiver partitions the present state of the game into two substates, and asks the source which substate contains the message. The (noisy) answer constitutes a vote against one substate or the other. The next state of the game is then a new list of numbers of words having received various numbers of negative votes. The general situation is depicted below:

n-state	partition	resulting (n-1)-state if answer favors left	resulting (n-1)-state if answer favors right
c_4	a_4 b_4	a_4	b_4
c_3	a_3 b_3	$a_3 + b_4$	$a_4 + b_3$
c_2	a_2 b_2	$a_2 + b_3$	$a_3 + b_2$
c_1	a_1 b_1	$a_1 + b_2$	$a_2 + b_1$
c_0	a_0 b_0	$a_0 + b_1$	$a_1 + b_0$
	$a_i + b_i = c_i$	$c_i' = a_i + b_{i+1}$	$c_i' = a_{i+1} + b_i$

It is frequently more convenient to discuss only the current state and the pair of states which may result from it, without being too concerned with the details of the partition which brings this about. This nonchalantness is justified by the following theorem:

Partitioning Theorem 2.1: There exists a partition which reduces the state $\underline{x}$ into the states $\underline{y}$ and $\underline{z}$ iff

(1) $\underline{x} \geq 0$; $\underline{y} \geq 0$; $\underline{z} \geq 0$ (all components are nonnegative)

(2) $x_{i+1} + x_i = y_i + z_i$ for all i, $0 \leq i \leq e$

(3) For all I, $0 \leq I \leq e$

$$\sum_{i=I}^{e} y_{2i+2} \leq \sum_{i=I}^{e} z_{2i+1} \leq \sum_{i=I}^{e} y_{2i}$$

and

$$\sum_{i=I}^{e} z_{2i+2} \leq \sum_{i+I}^{e} y_{2i+1} \leq \sum_{i=I}^{e} z_{2i}$$

For given $\underline{x}$, $\underline{y}$, and $\underline{z}$, this partition is unique.

Proof: Without (1), the vectors $\underline{x}$, $\underline{y}$, and $\underline{z}$ are not really states and partitioning is meaningless. Among nonnegative vectors, a partition exists if there are two substates $\underline{u}$ and $\underline{v}$ such that

$$\underline{x} = \underline{u} + \underline{v}$$
$$y_i = u_i + v_{i+1};\ z_i = v_i + u_{i+1} \quad \text{for all } i \ .$$

We will show that given $\underline{y}$ and $\underline{z}$, both $\underline{x}$ and the unique partition can be determined subject to conditions (2) and (3). Solving for $\underline{x}$ is most readily accomplished by computing the highest components first and working down.

$$x_e = u_e + v_e = y_e + z_e$$

In general, $x_i + x_{i+1} = u_i + u_{i+1} + v_i + v_{i+1} = y_i + z_i$, and thus x can be computed from the topmost component working down, using the equation $x_i = y_i + z_i - x_{i+1}$.

We may also solve for $\underline{u}$ and $\underline{v}$ in terms of $\underline{y}$ and $\underline{z}$.

$$\sum_{i=I}^{e} y_{2i+1} = \sum_{i=I}^{e} (u_{2i+1} + v_{2i+2});\ \sum_{i=I}^{e} z_{2i} = \sum_{i=I}^{e} (v_{2i} + u_{2i+1})$$

$$v_{2I} = \sum_{i=I}^{e} (v_{2i} - v_{2i+2}) = \sum_{i=I}^{e} (z_{2i} - y_{2i+1}) \ .$$

Similar expressions are found for the odd components of v, and for the odd and even components of $\underline{u}$. Condition (3) is the statement that these components be nonnegative. Q.E.D.

For some n-states, it is possible to devise a partitioning strategy for the remaining n questions which ensures that all words but one will eventually receive more than e negative votes; for other n-states no such strategy exists. We call the former winning n-states; the latter, losing n-states. A 0-state is winning if only one word has e or less negative votes. These considerations justify the following definitions:

Definitions 2.2: A 0-state $\underline{x}$ is winning if $\Sigma x_i \leq 1$. A nonzero winning 0-state is called a singlet. An n-state is winning if it can be reduced to two winning (n-1)-states. (The two winning (n-1)-states need not be distinct.)

Any vector which is a winning j-state but a losing (j-1)-state is said to be a borderline winning j-state. Several lemmas follow at once:

Lemma 2.3: Any vector which is a winning n-state is also a winning j-state, for any $j > n$. Singlet states are winning n-states for all n.

Lemma 2.4: The only borderline winning 1-state is

$$\begin{matrix} 0 \\ \cdots \\ 0 \\ 2 \end{matrix} .$$

Omitting top zeros, this state is written as 2.

If $\Sigma x_i = 2$, $\underline{x}$ is called a doublet. The winning partition of any doublet (if it exists) plays the two words against each other. This consideration leads to the following result:

Lemma 2.5: A doublet $\underline{c}$ is a winning n-state iff $\Sigma ic_i \leq n-1$, with equality in the borderline case.

Lemma 2.6: If $\underline{c} \leq \underline{d}$ (meaning $c_i \leq d_i$ for all i) and $\underline{d}$ is a winning n-state, then $\underline{c}$ is also a winning n-state.

The conclusion of Lemma 2.6 is also valid under slightly weaker hypotheses:

Lemma 2.7: If $\sum_{i=k}^{e} c_i \leq \sum_{i=k}^{e} d_i$ for all k, and $\underline{d}$ is a winning n-state, then $\underline{c}$ is also a winning n-state.

Lemma 2.8: M is a winning n-state if $M \leq 2^n$. The best partition of the state M is one which plays half the words against the other half.

A table of some of the winning n-states for $1 \leq n \leq 9$ is given in Figure 6.

3. Upper Bounds on Error-Correction Capability

An examination of Figure 6 leads us to some more general results. Foremost among these is a volume bound, which is a generalization of Hamming's (1950) bound for one-way codes. The primary difference is that our bound surrounds words at different levels with different sizes of spheres.

The appropriate definition of the volume of an n-state $\underline{x}$ is obtained by surrounding all words at height j by a sphere of radius j:

$$V_n(x) = \sum_{i=0}^{e} x_i \sum_{j=0}^{i} \binom{n}{j}$$

Theorem 3.1 (Conservation of Volume): Let $\underline{x}$ be any particular n-state, and let $\underline{y}$ and $\underline{z}$ be the (n-1)-states which result from it following any given partition. Then $V_n(\underline{x}) = V_{n-1}(\underline{y}) + V_{n-1}(\underline{z})$.

Proof: Let $\underline{x} = \underline{u} + \underline{v}$ be the partition which reduces $\underline{x}$ to $\underline{y}$ and $\underline{z}$. Then the theorem becomes

$$\sum_{i=0}^{e} (u_i + v_i) \sum_{j=0}^{i} \binom{n}{j} = \sum_{i=0}^{e} (u_i + v_i + u_{i+1} + v_{i+1}) \sum_{j=0}^{i} \binom{n-1}{j} .$$

Figure 6

SOME WINNING n-STATES, $1 \leq n \leq 9$

i n:	9	8	7	6	5	4	3	2	1
0	512	256	128	64	32	16	8	4	2
1	50	28	16	8	4	2	2	1	
0	12	4	0	8	8	6	0	1	
2	1	1	1	1	1	1	1		
1	0	0	0	0	0	0	0		
0	456	219	99	42	16	5	1		
2	1	1	1	1	1	1			
1	43	22	11	4	1	1			
0	36	21	11	14	10	0			
.	.	2	2	2	2				
.	.	0	0	0	0				
.	.	182	70	20	0				
	.	2	2						
	.	20	6						
	.	2	22						
	7	3							
	0	0							
	190	145							
	7	3							
	15	14							
	40	19							
	8	4							
	0	0							
	144	108							
	8	4							
	8	8							
	64	36							
	1	1	1	1	1	1			
	4	1	1	0	0	0			
	14	10	0	1	1	0			
	58	36	35	15	0	1			
	.	.	1	1	1				
	.	.	0	0	0				
			5	0	0				
			24	22	6				

Since x_i is arbitrary, an equivalent theorem is

$$\sum_{j=0}^{i} \binom{n}{j} = \sum_{j=0}^{i} \binom{n-1}{j} + \sum_{j=0}^{i-1} \binom{n-1}{j} .$$

This is true because

$$\binom{n}{j} = \binom{n-1}{j} + \binom{n-1}{j-1} \qquad \text{Q.E.D.}$$

One immediate application of this result is Theorem 3.2.

Volume Bound Theorem 3.2: If $\underline{x}$ is a winning n-state, then

$$V_n(\underline{x}) \leq 2^n .$$

Proof: The theorem is true for $n = 0$, for in fact a singlet state satisfies any volume bound:

$$\sum_{k=0}^{\infty} \binom{j}{k} = \sum_{k=0}^{j} \binom{j}{k} = 2^j$$

For arbitrary n, the theorem follows directly from the conservation of volume theorem by induction. Q.E.D.

In some special cases this bound is the only requirement. Lemma 2.8 showed one such case. Doublet states are another, as is shown by the following restatement of Lemma 2.5.

Lemma 3.3: A doublet state $\underline{c}$ is a winning n-state iff $V_n(\underline{c}) \leq 2^n$. Equality occurs in the borderline case.

Proof: Let the only nonzero components of $\underline{c}$ be $c_i = 1 = c_j$ (where possibly $i = j$). Then the borderline case of Lemma 2.5 becomes $i + j = n - 1$. In this case

$$V_n(\underline{c}) = \sum_{k=0}^{i} \binom{n}{k} + \sum_{k=0}^{j} \binom{n}{k} = \sum_{k=0}^{i} \binom{n}{k} + \sum_{k=n-j}^{n} \binom{n}{n-k} = \sum_{k=0}^{n} \binom{n}{k} = 2^n$$

Q.E.D.

Lemma 3.3 shows that for doublets, the volume bound is the only restriction that must be satisfied. In general, however, there are losing states which still satisfy the volume bound. For example, consider the 4-state $\begin{smallmatrix}3\\0\end{smallmatrix}$. Its volume is $3(\binom{4}{0} + \binom{4}{1}) = 15 < 16 = 2^4$, but this is nevertheless a losing state. A generalized version of this limitation is the following.

Translation Bound Theorem 3.4: If $\Sigma x_i \geq 3$, then $\underline{x}$ cannot be a winning n-state unless $T\underline{x}$ is a winning (n-3)-state. $T\underline{x}$ is the translation of $\underline{x}$, defined by $(T\underline{x})_i = x_{i+1}$.

Proof: The basic idea is again an induction on n . We first verify by exhaustion that the theorem is true for small n, as shown in Figure 6. Now suppose that the theorem is true for $n \leq k - 1$, and that $\underline{x}$ is a winning k-state. There must then exist some partition of $\underline{x}$ which reduces it to $\underline{y}$ and $\underline{z}$, which are winning (k-1)-states. When this same partition is applied to $T\underline{x}$, it reduces $T\underline{x}$ to $T\underline{y}$ and $T\underline{z}$. If $\Sigma y_i \geq 3$ and $\Sigma z_i \geq 3$, then the induction hypothesis guarantees that $T\underline{y}$ and $T\underline{z}$ are both winning (k-r)-states. Thus, $T\underline{x}$ is a winning (k-3)-state.

In the exceptional case that $\Sigma y_i \leq 2$, we must resort to special considerations. The translation bound, as stated, does not apply to such states. In fact, from Lemma 2.5, if the doublet $\underline{y}$ is a borderline winning n-state, then $T\underline{y}$ is a borderline winning (n-2)-state. If $\underline{y}$ is not a borderline (k-1)-state, then $T\underline{y}$ is a winning (k-3)-state. Thus only the borderline doublet must be considered.

Define $\underline{x}'$ by $x'_0 = 0$; $x'_k = x_k$ for all $k > 0$. Then $\underline{x}' \leq \underline{x}$ and $V_n(\underline{x}') \leq V_n(\underline{x})$. Since $\underline{y}$ is a reduction from $\underline{x}$,

$$\sum_{i=0}^{e} y_i \geq \sum_{i=1}^{e} x_i = \sum_{i=0}^{e} x'_i \ .$$

Thus $\underline{x}'$ is a singlet or a doublet. If it is a singlet, so is $T\underline{x} = T\underline{x}'$, and the proof is completed. If $\underline{x}'$ is a doublet, there are two possibilities. Either $\underline{x}' = \underline{y}$, or $\underline{x}'$ is a borderline winning k-state (because the doublet $\underline{x}$ reduces to $\underline{y}$ and $\underline{y}$ is a borderline winning (k-1)-state).

If $\underline{x}' = \underline{y}$ then $T\underline{x} = T\underline{y}$ and $T\underline{x}$ is a winning (k-3)-state because it is a translate of the winning doublet (k-1)-state.

If instead, $\underline{x}'$ is a borderline winning k-state, then it satisfies the volume bound with equality: $V_k(\underline{x}') = 2^k$. But x is also a winning k-state. $\underline{x} = \underline{x}'$ is the only possibility. In this case $T\underline{x}$ is not a winning (k-3)-state, but $\underline{x}$ is a doublet, so this case lies outside the hypotheses of the theorem.

Having verified the theorem in all the exceptional cases, we have completed the proof. Q.E.D.

Corollary 3.5: If $x_e \geq 3$, then $\underline{x}$ cannot be a winning n-state unless $n \geq 3e + 2$.

Together, the volume bound and the translation bound eliminate most of the losing states. They are not exhaustive, however, for there are a few losing states which satisfy both bounds. For example, $\frac{51}{2}$ is a losing 9-state, even though it satisfies the volume bound and 51 is a winning 6-state. In spite of such isolated cases, however, we shall show in Part 4 that in the asymptotic cases of greatest interest, these two bounds are all inclusive.

The volume and translation bounds apply to all possible n-states. The special n-states of greatest interest are ones in which the initial state $\underline{I}$ contains only one nonzero component: $I_e = M = 2^k$. We wish to find the minimum n for which $\underline{I}$ is a winning n-state. The information rate is then defined by $R = k/n$; the allowable error fraction is denoted by $f = e/n$. For small values of n, the possible values of R and F can be derived from our table. We note, for example, the occurrence of $\begin{smallmatrix}16\\0\end{smallmatrix}$ among our list of winning 7-states. (There is actually a one-way partitioning strategy which accomplishes this win, called the Hamming (7, 4) single-error-correcting code.) For $n > 9$, however, the detailed extension of the table involves considerable labor and we must rely instead on asymptotic bounds which we shall now derive. We are interested in the cases when n grows very large while f and R remain fixed.

3.6. The Asymptotic Volume Bound and Its Tangents: Let $f = e/N$. We first consider the volume bound:

$$2^k \sum_{j=0}^{e} \binom{n}{j} \leq 2^n .$$

Asymptotically this becomes $H(f) \leq 1 - R$, or $R \leq 1 - H(f)$, where $H(f) = -f \ln f - (1-f) \ln(1-f)$. This is identical to Hamming's bound for one-way codes, as plotted in Figures 1 and 2.

In Part 4 of this paper, the tangents to this curve play an important role. We digress here briefly to derive the equations for these tangent lines.

Consider a straight line which intercepts the f axis at f_0, the R axis at R_0, and which is tangent to the curve $R = 1 - H(f)$ at the point $R = R_t$, $f = f_t$. Define $g = 1 - f$. Then the equations giving the point of tangency are

$$R(f_t) = 1 - H(f_t) = R_0(1 - f_t/f_0) \qquad (3.7)$$

$$R'(f_t) = \log(f_t/g_t) = -R_0/f_0 \qquad (3.8)$$

and

$$f_t + g_t = 1 .$$

These quantities are depicted in Figure 7. Subtracting f_t times Equation 3.8 from Equation 3.7 gives

$$R_0 = 1 + \log g_t$$

Figure 7

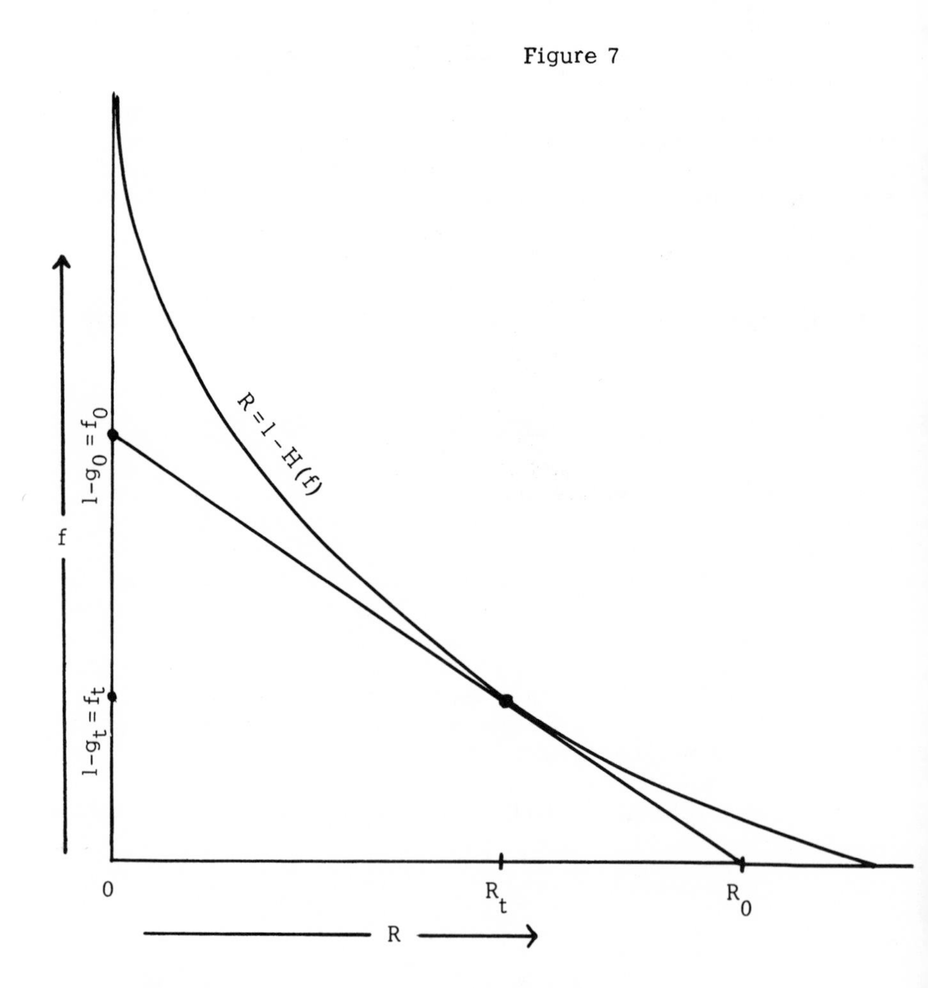

Substituting this expression into Equation 3.8 and exponentiating gives

$$2f_t^{f_0} g_t^{g_0} = 1 \ .$$

Introducing the quantity $s = g_t/f_t$, we then have $\log s = R_0/f_0$; $f_t = 1/(1+s)$; $R_t = (f_0 - 1/(1+s)) \log s$. These substitutions transform the problem into a single equation in only one unknown:

$$1 = 2f_t^{f_0} g_t^{g_0} = 2s^{g_0}/(1+s)$$

or

$$2s^{g_0} = 1 + s \ . \tag{3.9}$$

In the special case in which g_0 is a rational number, this equation is algebraic. $s = 1$ is always an extraneous root of this equation; it may be removed by dividing through by $s - 1$. The computation of the coordinates of the tangency point then reduces to the solution of this final algebraic equation.

Example 3.10: $g_0 = 2/3$. In this case the equation is

$$2s^{2/3} = 1 + s$$

$$8s^2 = s^3 + 3s^2 + 3s + 1$$

$$0 = s^3 - 5s^2 + 3s + 1 = (s-1)(s^2 - 4s - 1)$$

$$s = 2 + 5^{\frac{1}{2}} = ((1 + 5^{\frac{1}{2}})/2)^3$$

$$f_t = (3 + 5^{\frac{1}{2}}) = .19095$$

$$R_t = (1/3 - 1/(3 + 5^{\frac{1}{2}})) \log(2 + 5^{\frac{1}{2}}) = .29650$$

$$R_0 = \log((1 + 5^{\frac{1}{2}})/2) = .69425 \ .$$

3.11. The Tangential Bound: The best possible asymptotic upper bound on error-correction capability is obtained by the following argument:

We first apply the translation theorem: If the initial state $\underline{I}$ (which has 2^k words at height e) is a winning n-state, then $T^{\overline{m}}$ $\underline{I}$ must be a winning $(n - 3m)$-state, for any $0 < m < n/3$. Applying the volume bound to this state gives

$$2^k \binom{n-3m}{e-m} \leq 2^{n-3m} \quad .$$

Define $x = n - 3e$; $y = e - m$. The bound becomes

$$8^{-y} \binom{x+3y}{y} \leq 2^{x-k}$$

The validity of this bound is restricted only by the requirement that $0 < y < e$, and it behooves us to choose the best y to obtain the strongest bound. This is accomplished by maximizing the left side of the above inequality. This can be done most readily by setting equal to one the ratio of the value of this expression for y to its value for $y + 1$. For large y, this gives

$$1 = \frac{(x+3y)^3}{8y(x+2y)^2}$$

$$0 = (x+y)(x^2 - 5y^2)$$

$$y = 5^{-\frac{1}{2}} x \quad .$$

Plugging this value into the bound and taking logarithms gives

$$x(1 + 3 \cdot 5^{-\frac{1}{2}}) H(1/(3 + 5^{\frac{1}{2}})) \leq x(1 + 3 \cdot 5^{-\frac{1}{2}}) - k$$

or

$$R(1 + 3 \cdot 5^{-\frac{1}{2}})^{-1} \leq (1 - 3f)(1 - H(1/(3 + 5^{\frac{1}{2}}))) \quad . \qquad (3.12)$$

The result is valid in the region $0 < y < e$, which is equivalent to the requirement that

$$(3 + 5^{\frac{1}{2}})^{-1} < f < 1/3 \quad .$$

Comparing these numbers with the tangents to the volume bound computed in Example 3.10, we see that this bound is a straight line which goes from the point $R = 0$, $f = 1/3$ to the volume bound, where it comes in tangentially and then ends. A plot of this bound is given in Figure 2.

This bound is a special case of a more general bound for error exponents given by Shannon, Gallager, and Berlekamp (1967).

In Part 4, we shall show how this bound is actually attainable.

4. The Construction of Asymptotically Optimum Coding Strategies

Having completed proofs of the volume bound, the translation bound, and their asymptotic combination, we are naturally led to investigate the possibility of finding specific winning states which lie on or close to these bounds. We start by an examination of our table of winning states for small n, (Figure 6). We know that in order to find any substantial (≥ 3) number of words at the top component, we are restricted to states for which $n \geq 3e + 2$. Thus if $e = 0$, $n \geq 2$, and we find that 4 is indeed a winning 2-state. If $e = 1$, $n \geq 5$, and we find that $\begin{matrix}4\\8\end{matrix}$ is a winning 5-state. Continuing, we find that $\begin{matrix}4\\8\\36\end{matrix}$ is a winning 8-state. This is quite a bit better than we had bargained for!! We knew that $\begin{matrix}3\\0\end{matrix}$ is a losing 7-state, and were inquiring merely as to whether it is a winning 8-state. We find that not only can we put $4 (>3)$ words on top, but a sizable number of additional words may be added at the lower levels. If we continue this investigation, we find that $\begin{matrix}4\\8\\36\\152\end{matrix}$ is a winning 11-state. Further extensions of this sequence are found in the first column of the table of Figure 8.

Figure 8

INFINITE SEQUENCE OF BORDERLINE WINNING STATES

Row \ Column:	1	2	3	4	5	6	7	8	9	10	11	12	13	14
1	4	2	1	1	1	1	1	1	1	1	1	1	1	1
2	8	6	4	1	0	0	0	0	0	0	0	0	0	0
3	36	22	14	10	5	1	0	0	0	0	0	0	0	0
4	152	94	58	36	24	15	6	1	0	0	0	0	0	0
5	644	398	246	152	94	60	39	21	7	1	0	0	0	0
6	2728	1686	1042	644	398	246	154	99	60	28	8	1	0	0
7	11556	7142	4414	2728	1686	1042	644	400	253	159	88	36	9	1
⋮														

MOST IMPORTANT PROPERTIES

Let $A_{i,j}$ be the number in the ith row and the jth column:

$$\underline{A}_{m,j} = \begin{matrix} A_{1,j} \\ A_{2,j} \\ \cdots \\ A_{m,j} \end{matrix}$$

$\underline{A}_{m,j}$ is a borderline winning $(3m-j)$-state. It satisfies volume bound with equality. It can be reduced to $\underline{A}_{m,(j+1)}$ and $\underline{A}_{(m-1),(j-2)}$

$$A_{i,j} + A_{i,(j+1)} = A_{(i+1),(j+2)} \quad \text{(unless } i \leq 2)$$

If $i \geq j$ and $i \geq 3$, then

$$A_{i,j} = 2((1+5^{\frac{1}{2}})/2)^{3i-j-2} + 2((1-5^{\frac{1}{2}})/2)^{3i-j-2}$$

Figure 9

ANOTHER INFINITE SEQUENCE OF BORDERLINE WINNING STATES

8	4	2	1	1	1	1	1	1	1	1
64	36	20	11	4	1	0	0	0	0	0
744	404	220	120	67	35	16	5	1	0	0
8512	4628	2516	1368	744	407	222	118	22	6	1

We shall now construct this table and derive some of its important properties.

Definition 4.1: The values in the table of Figure 8 are defin recursively as follows: The first two rows are postulated as initial conditions:

$$A_{1,1} = 4; A_{1,2} = 2; A_{1,k} = 1 \quad \text{for} \quad k \geq 3$$

$$A_{2,1} = 8; A_{2,2} = 6; A_{2,3} = 4; A_{2,4} = 1; A_{2,k} = 0 \quad \text{for} \quad k \geq 5 \ .$$

The remainder of the table is derived recursively by the following rules; applicable only when $i \geq 3$.

For $j \geq 3$, $A_{i,j} = A_{i-1,j-1} + A_{i-1,j-2}$

For $j = 2$, $A_{i,2} = A_{i,3} + A_{i-1,1}$

For $j = 1$, $A_{i,1} = A_{i,2} + A_{i,3}$.

Definition 4.2: The state $\underline{A}_{m,j} = \begin{matrix} A_{1,j} \\ A_{2,j} \\ \cdots \\ A_{m,j} \end{matrix}$

Theorem 4.3: For $j \leq 3 \leq i$, $A_{i,j} = 2((1+5^{\frac{1}{2}})/2)^{3i-j-2} + 2((1-5^{\frac{1}{2}})/2)^{3i-j-2}$.

Proof: Notice that the first three columns are defined only in terms of themselves. We introduce the single sequence a_k by the transformation:

$$a_k = A_{(k+3)/3,1} \quad \text{if } k \equiv 0 \bmod 3$$

$$a_k = A_{(k+4)/3,2} \quad \text{if } k \equiv 2 \bmod 3$$

$$a_k = A_{(k+5)/3,3} \quad \text{if } k \equiv 1 \bmod 3 \ .$$

The recurrence relations defining $A_{i,j}$ then become

$$a_k = a_{k-1} + a_{k-2}, \quad \text{valid for } k \geq 4 \ .$$

The general solution of this equation is of the form

$$a_k = B\, r_1^k + C\, r_2^k$$

where B and C are constants determined by the two initial conditions, $a_2 = 6$ and $a_3 = 8$. r_1 and r_2 are the roots of the equation

$$r^2 = r + 1 \ .$$

Solving gives $r = (1 \pm 5^{\frac{1}{2}})/2$, $B = C = 2$. Transforming back from the a_k to the $A_{i,j}$ gives the desired result. Q.E.D.

Corollary 4.4: Theorem 4.3 also holds in the extended range $i \geq j$, $i \geq 3$.

Proof: These values are obtained by the same recurrence relations as their counterparts in the first three columns, to which they must be equal.

Theorem 4.5: $\underline{A}_{m,j}$ can be reduced to $\underline{A}_{m-1,j-2}$ and $\underline{A}_{m,j+1}$.

Proof: We first patch up the exceptional columns on the left boundary, for $j \leq 2$, by defining $A_{m-1,0} = A_{m,3}$; $A_{m-2,-1} = A_{m,2}$. The recurrence relations defining the table are than uniformly stated for all $m \geq 0$. The proof consists of verifying conditions (2) and (3) of the partitioning Theorem 2.1. Condition (2) becomes

$$A_{i,j} + A_{i-1,j} = A_{i,j+1} + A_{i-1,j-2} \ .$$

If $i = 1$ or 2, we observe that this condition is satisfied by the initial conditions. For larger i, we have

$$A_{i,j} = A_{i-1,j-1} + A_{i-1,j-2}$$

$$A_{i,j+1} = A_{i-1,j} + A_{i-1,j-1} \ .$$

Subtraction of these two equations yields condition (2). A sufficient condition for satisfying (3) is

$$y_{2i+2} \leq z_{2i+1} \leq y_{2i} \quad \text{and} \quad z_{2i+2} < y_{2i+1} \leq z_{2i}$$

$$A_{i-2,j-2} \leq A_{i,j+1} \leq A_{i,j+2} \ .$$

The latter inequality follows from the fact that $A_{i,j}$ is a monotonic nonincreasing function of j, for any fixed i. This monotonicity may be established by induction on i and the observation that the monotonicity holds for $j \leq 3$, where $A_{i,j}$ is given by an explicit formula. The former inequality is verified as follows:

$$A_{i,j+1} = A_{i-1,j} + A_{i-1,j-1} \geq A_{i-1,j} = A_{i-2,j-1}$$

$$+ A_{i-2,j-2} \geq A_{i-2,j-2}$$

Q.E.D.

Corollary 4.6: $\underline{A}_{m,j}$ is a winning $(3m-j)$-state. (proof by induction)

Corollary 4.7: $V_{3m-j}(\underline{A}_{m,j}) = 2^{3m-j}$ (proof by induction, using conservation of volume Theorem 3.1).

This concludes our proof of the remarkable properties of the table of Figure 8.

The correct partition of any state shown in column 1 or 2 of Figure 8 divides the words at each level evenly between the two sets. The states shown in the other columns cannot be so partitioned, because they have a single word at the top. To balance the two sets, certain numbers of "compensating words" at various lower levels must be assigned to the other set. It turns out that the correct numbers of compensating words needed to partition the jth column correctly are

$$0$$

$$\binom{0}{j-3} + \binom{1}{j-3}$$

$$\binom{1}{j-4} + \binom{2}{j-4}$$

$$\binom{2}{j-5} + \binom{3}{j-5}$$

$$\vdots$$

For example, the correct partitioning of the 6th column is

$$\begin{matrix} 1 \\ 0 \\ 1 \\ 15 \\ 60 \\ 246 \\ 1042 \\ \vdots \end{matrix} = \begin{matrix} 1 + & 0 \\ & 0 \\ & 0 \\ & 5 \\ & 29 \\ & 123 \\ & 521 \\ & \vdots \end{matrix} + \begin{matrix} 0 + 0 \\ 0 + 0 \\ 1 + 0 \\ 5 + 5 \\ 2 + 29 \\ 0 + 123 \\ 0 + 521 \\ \vdots \end{matrix}$$

We can use the first column of Table 8 to obtain a lower bound on f and R for large n. The manipulations are simple:

$$2^k \leq 2[(1+5^{\frac{1}{2}})/2]^{3(j-1)} = A_{j,1} \quad \text{(to the nearest integer)}$$

so if

$$k - 1 \leq (j-1)\, 3 \log[(1+5^{\frac{1}{2}})/2]$$

we have

$$I \leq A_{e+j,1}$$

which is a winning $3(e+j-1)$-state. Thus we may protect 2^k words from e errors by n questions if

$$k - 1 \leq (n-3e) \log((1+5^{\frac{1}{2}})/2)$$

$$R \leq (1-3f) R_0 ; \; R_0 = \log((1+5^{\frac{1}{2}})/2) \;. \qquad (4.8)$$

Since Equation 4.8 is identical to Equation 3.12, we conclude that for all rates in the region $0 < R < R_t = .30$, (or equivalently, for all error fractions $.19 = f_t < f = e/n \leq f_0 = 1/3$) this straight line gives the best possible asymptotic result. For higher rates (or lower error fractions), however, the bounds differ. The upper bound on R and f departs from the straight line and follows the curve $1-H(f)$. The lower bound obtained by this simple comparison with the first column of Figure 8 remains on the straight line for small error fractions, f.

We soon decide that in the region of this discrepancy it is the straight line, rather than the volume curve, which is the weak bound. Among other observations, we know from Gilbert's (1952) bound for one-way codes that it is possible to get the rate R up to 1 for sufficiently small error fractions f. We are thus led to search for sequences of winning states which provide better achievable bounds for high rates.

In this region the restraining limits are imposed by the volume bound rather than the translation bound. Intuition suggests that we would do well to construct infinite sequences of winning states which build up more slowly, having fewer nonzero components, but much greater weight. These bottommore components must carry the load for small e/n. These desires can be fulfilled by requiring that the translate of a borderline winning n-state be a borderline winning $(n-4)$-state instead of an $(n-3)$-state. The attempt to construct a table with this property proves successful. The result is shown in Figure 9.

In fact, for any integer $t \geq 3$, we can compute a similar table of winning states. The first two lines of this table are defined as

$$\begin{array}{l} 2^{t-1} \ldots 2^{t-u} \qquad \ldots 1 \quad \ldots \quad 1 \quad \ldots 11111 \quad \ldots \\ 2^{t-1}(2^t-2t) \ldots 2^{t-u}(2^t-2t\tfrac{1}{2}u-1) \ldots 2^t-t-1 \ldots 2^v-v-1 \ldots 11\ 4\ 1\ 0\ 0 \quad \ldots \end{array}$$

The remainder of the table is recursively defined to fulfill the relation

$$A_{i,j} + A_{i-1,j} = A_{i,j+1} + A_{i-1,j-(t-1)} .$$

This definition cinches condition (2) of the partitioning Theorem 2.1. Condition (3) of the partitioning theorem is readily verified by methods similar to those elucidated for the special case $t = 3$. This then proves that the constructed table has the basic property:

$$\underline{A}_{m,j} \text{ reduces to } \underline{A}_{m,j+1} \text{ and } \underline{A}_{m-1,j-(t-1)} .$$

To compute the rate of exponential growth of the components of the first column of this table, we can again change variables to a_k. The first t columns then become

$$\begin{array}{llll} a_t & a_{t-1} & \cdots & a_1 \\ a_{2t} & a_{2t-1} & \cdots & a_{t+1} \\ a_{3t} & a_{3t-1} & \cdots & a_{2t+1} . \end{array}$$

In terms of the a_k, the recursion relation becomes

$$a_k + a_{k-1} = 2a_{k-1}$$

from which we get an algebraic equation for the growth rate r

$$r^t + 1 = 2r^{t-1} \quad .$$

The growth rate of the first column, $s = r^t$, satisfies the equation

$$s + 1 = 2s^{(t-1)/t} \quad . \tag{4.9}$$

In terms of s, the achievable asymptote becomes

$$s^k \leq B\, s^j; \; j + e \leq nt$$

$$k \leq (nt-e) \log s$$

$$R \leq (t-f) \log s$$

Comparing Equation (4.9) with Equation (3.9), we note that if $g_0 = (t-1)/t$, the equations are identical!! This bound is a straight line, emanating from the point $f_0 = 1/t$ (or equivalently, $g_0 = (t-1)/t$) and proceeding up to a point where it touches the volume bound tangentially, and then continues on to the R axis. For lower and lower rates, the best bounds result from higher and higher values of the integer t .

There is considerable heuristic evidence that the correct asymptotic relationship between f and R is given by the volume bound for all rates $R \geq .2965 \ldots$. However, the only known lower bounds on f and R are based on tables whose constructions require that the number t in Equation (4.9) be an integer. It is conjectured that similar tables may be constructed in which t is any rational number ≥ 3 . An unproved construction with $t = 7/2$ is tentatively presented in Tables 10 and 11.

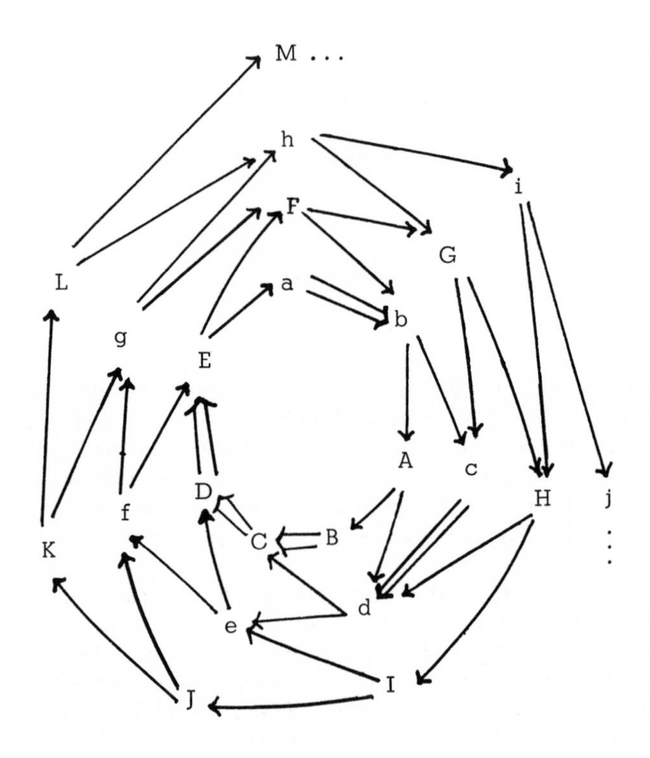

Figure 10. Reduction transition diagram for states of Figure 11.

A	B	C	D	E	F	G	H	I	J	K	L	M	N	O	P	Q	R
1																	
11	8	4	2	1	1	1	1	1	1	1	1	1	1	1	1	1	1
57	32	20	12	7	2	0	0	0	0	0	0	0	0	0	0	0	0
395	224	128	74	43	28	16	5	1	0	0	0	0	0	0	0	0	0
3385	1912	1068	598	336	183	102	61	30	11	2	0	0	0	0	0	0	0
16429	9760	5836	3452	2025	1205	712	419	252	146	75	29	7	1	0	0	0	0
						5810	3176	1727	951	531	294	146	56	15	2	0	0
						27996	17044	11166	6807	3096	1424	1408	779	382	154	46	9

a	b	c	d	e	f	g	h	i	j	k	l	m
6	3	2	1	1	1	1	1	1	1	1	1	1
22	14	6	4	1	0	0	0	0	0	0	0	0
196	109	66	36	20	9	2	0	0	0	0	0	0
1156	676	390	228	136	82	48	22	6	1	0	0	0
10954	6055	3346	1868	1028	566	312	177	97	42	13	2	0
43318	27136	16762	10054	6086	3662	2203	1310	775	453	243	110	37

Figure 11. Another infinite sequence of winning states?

REFERENCES

Berlekamp, E. R., 1964, "Block Coding with Noiseless Feedback," Ph. D. thesis, Department of Electrical Engineering, Massachusetts Institute of Technology, Cambridge, Massachusetts.

Berlekamp, E. R., 1968, "Algebraic Coding Theory," McGraw-Hill Book Co., New York.

Gilbert, E. N., 1952, A Comparison of Signaling Alphabets Bell System Tech. J., 31: 504-522.

Hamming, R. W., 1950, Error Detecting and Error Correcting Codes Bell System Tech. J., 29: 147-160.

Horstein, M., 1963, Sequential Transmission Using Noiseless Feedback, IEEE Trans. Inform. Theory, IT9: 136-143.

Schalkwijk, J. P. M., and T. Kailath, 1966, A Coding Scheme for Additive Noise Channels with Feedback, IEEE Trans. Inform. Theory, IT12: 172-189.

Shannon, C. E., 1956, Zero-Error Capacity of Noisy Channels IEEE Trans. Inform. Theory, IT2: 8-19.

Shannon, C. E., R. G. Gallager, and E. R. Berlekamp, 1967, Lower Bounds to Error Probability for Coding on Discrete Memoryless Channels, Inform. Control, 10: 65-103, 522-552.

JAMES L. MASSEY

Some Algebraic and Distance Properties of Convolutional Codes

ABSTRACT

A Plotkin-type upper bound and a Gilbert-type lower bound are proved for the feedback-decoding and definite-decoding minimum distances respectively of binary convolutional codes. In the former case, a very simple bound is derived which shows that the ratio of feedback decoding minimum distance to constraint length is asymptotically upper-bounded by $\frac{1}{2}(1-R)$ for codes of rate R. In the latter case, it is shown that the ratio d_{DD}/n_{DD} of definite-decoding minimum distance to constraint length is asymptotically lower bounded as $H(d_{DD}/n_{DD}) \geq 0.1\,(1-R)/(1+R)$ for the best codes of rate R. This derivation requires the development of several interesting relationships between convolutional codes and linear feedback shift-registers.

1. Introduction

In the following sections, we shall present some bounds on the attainable minimum distance for convolutional codes. The derivation of these bound leads also to the establishment of certain interesting algebraic properties of these codes. The results herein have evolved over the past several years and have benefitted considerably from the work of several of our students, particularly R. Kolor and W. Wilder, to the point that they are now sufficiently unified to offer to other workers in the field. In this section, we review the necessary background material on convolutional codes required in the following sections.

In this paper, only convolutional codes in canonic systematic form will be considered and the reader is referred to the literature [1] for the justification that this results in no loss of generality. For convenience, the discussion will also be restricted to binary codes unless explicitly stated to the contrary.

Let $\underline{i}_u$ denote a K-dimensional column vector over GF(2), the binary number field. Then $\underline{i}_0, \underline{i}_1, \ldots, \underline{i}_u, \ldots$ denotes a sequence of such vectors and we shall consider that the components of $\underline{i}_u$ are the K information digits to be encoded at time instant u . A (canonic systematic) convolutional code of memory order m is specified by the matrices $G_0, G_1, \ldots, G_m$ where each G_j is an $(N-K) \times K$ binary matrix. The N encoded digits at time instant u are the components of the column vector whose first K components form $\underline{i}_u$ and whose last $N-K$ components form the vector $\underline{p}_u$ given by

$$\underline{p}_u = G_0 \underline{i}_u + G_1 \underline{i}_{u-1} + \ldots + G_m \underline{i}_{u-m} \tag{1}$$

where we assume $\underline{i}_j = \underline{0}$ if $j < 0$. The components of $\underline{p}_u$ are the $N-K$ parity digits formed by the convolutional code at time instant u . The rate R of the code is defined as $R = K/N$.

It should be noted from (1) that a convolutional code of memory order $m = 0$ is just a systematic linear block code and conversely. Hence the theory of convolutional codes includes that of block codes as a special case which perhaps accounts for the greater difficulty with which distance bounds are derived for the former class of codes.

2. A Plotkin Upper Bound on Feedback Decoding Minimum Distance

The usual decoding method for convolutional codes, called feedback decoding (FD) by Robinson [2], calls for the decoding estimate of $\underline{i}_u$ to be made from the received digits over time units u through $u+m$ on the assumption that $\underline{i}_{u-1}, \ldots, \underline{i}_{u-m}$ have all been correctly decoded. With this assumption, the decoding of $\underline{i}_0$ is typical of the decoding of any $\underline{i}_u$ and hence the feedback-decoding minimum distance, d_{FD}, is appropriately defined as the fewest number of positions that two encoded sequences with differing values of $\underline{i}_0$ are found to disagree over the time span 0 through m . The total number of positions within this time span, namely $(m+1)N$, is called the feedback-decoding constraint length and will be denoted as n_{FD} . By the usual linearity argument, it follows that

$$d_{FD} = \min_{\underline{i}_0 \neq \underline{0}} W_H(\underline{i}_0, \underline{i}_1, \ldots, \underline{i}_m, \underline{p}_0, \underline{p}_1, \ldots, \underline{p}_m) \tag{2}$$

where $W_H(\ )$ denotes the Hamming weight, i.e. the number of nonzero components among the vectors, of the enclosed vectors.

In the remainder of this section, an upper bound on the ratio d_{FD}/n_{FD} will be obtained which as n_{FD} increases tends to the same value as the familiar asymptotic Plotkin upper bound for block codes with the same rate R . Hence this bound for convolutional codes will also be called a Plotkin bound. The derivation of this bound is facilitated by the introduction of the following:

Definition 1: The integer $d(m, N, K)$ is the maximum value of d_{FD} for the class of all convolutional codes of memory order m for encoding K binary digits per time instant into N digits.

In terms of the quantity $d(m, N, K)$, we next state four lemmas which will then be combined to yield the sought-for Plotkin bound.

Lemma 1: $d(m, N, K) \leq d(m, N-K+1, 1)$.

Proof: Note that the partial minimization of the righthand side of (2) over those $\underline{i}_u$ which are all zero in their last $K-1$ components is equivalent to the full minimization for the code with 1 information and $N - (K-1)$ encoded digits per time instant whose matrices G_j are just the first columns of the original matrices. Hence, for every code with K information and N encoded digits per time instant there is a code with 1 information and $N-K+1$ encoded digits per time instant whose minimum distance d_{FD} is at least as great as that of the first code.

Lemma 2: For N odd, $d(m, N, 1) \leq N + m\frac{N-1}{2}$.

Proof: We claim first that when $K = 1$, there will be a code having $d_{FD} = d(m, N, K)$ and having $G_0 = (1, 1, \ldots, 1)$ the $(N-1)$-dimensional all-one column vector. To show this, consider first any code in which the k-th component of G_0 is 0 . If this is also true for every G_j, then from (1) it follows that the k-th component of $\underline{p}_u$ is 0 for all u, so that changing the k-th component of G_0 to 1 will increase d_{FD} . Otherwise, let n be the smallest integer such that G_n has a 1 in the k-th component. From (1) it follows that $\underline{p}_u$ has a 0 in its k-th component for $u < n$. From (1) it also follows that moving the k-th component of G_{n+j} to the k-th component of G_j, $j = 0, 1, \ldots (m-n)$, has only the effect of moving the k-th component of $\underline{p}_{n+j}$ to the k-th component of $\underline{p}_j$ for all $j \geq 0$. Hence, we see from (2) that this new code has d_{FD} at least as great as the original code, and we also note that this new code has a 1 in the k-th component of G_0 . Since k is arbitrary, the claim follows.

It remains to show that when $G_0 = (1, 1, \ldots, 1)$ and $i_0 = 1$, then it is always possible to choose $i_1, \ldots, i_m$ to produce a vector $(i_0, i_1, \ldots, i_m, \underline{p}_0, \underline{p}_1, \ldots, \underline{p}_m)$ whose Hamming weight satisfies the inequality in the lemma. Note first that $(i_0, \underline{p}_0) = (1, G_0)$ and hence has weight N . But for any fixed code, any $u > 0$, and any fixed choice of $i_1, i_2, \ldots, i_{u-1}$, it follows from (1) that

$$(i_u, \underline{p}_u) = (i_u, i_u, \ldots, i_u) + (0, \sum_{j=1}^{u} G_j i_{u-j}) .$$

Since the second vector on the right is a fixed N-vector, it follows that the choices $i_u = 0$ and $i_u = 1$ result in N-vectors that are complements of one another for $(i_u, \underline{p}_u)$. But since N is odd, one of these N vectors must have weight at most $(N-1)/2$. Hence, for any code, we can choose $i_1, i_2, \ldots, i_m$ in order so that $(\underline{i}_u, \underline{p}_u)$ has weight at most $(N-1)/2$ for $u = 1, 2, \ldots, m$ and the lemma is proved.

Lemma 3: For m even, $d(m, N, K) \leq d(\frac{m}{2}, 2N, 2K)$.

Proof: For any code with parameters $m = 2m^*$, N, K and defining matrices $G_0, G_1, \ldots, G_{2m^*}$ we consider the new code with parameters m^*, $N^* = 2N$ and $K^* = 2K$ with defining matrices

$$G_j^* = \begin{bmatrix} G_{2j} & G_{2j-1} \\ G_{2j+1} & G_{2j} \end{bmatrix} \qquad j = 0, 1, \ldots, m^*$$

where we define G_{-1} and G_{m+1} both to be the all zero matrix. Also, we set the $N^* = 2N$ vector

$$\underline{i}_j^* = (\underline{i}_{2j}, \underline{i}_{2j+1})$$

and set the $N^* - K^* = 2(N-K)$ vector

$$\underline{p}_j^* = (\underline{p}_{2j}, \underline{p}_{2j+1}) \quad .$$

It is then readily checked that for this new code, p_u satisfies (1) with the matrices G_j of the old code. Hence, from (2), it follows that the minimum distance of the new code satisfies

$$d_{FD}^* = \min_{\substack{\underline{i}_0 \neq 0 \text{ or } \underline{i}_1 \neq 0 \\ \text{or both}}} W_H(\underline{i}_0, \underline{i}_1, \ldots, \underline{i}_{m+1}, \underline{p}_0, \underline{p}_1, \ldots, \underline{p}_{m+1})$$

where the righthand side is evaluated for the old code. Hence the minimum clearly occurs with $\underline{i}_0 = \underline{0}$ (which implies $\underline{p}_0 = \underline{0}$) and $\underline{i}_1 \neq \underline{0}$ so that the righthand side of the preceding equation differs from (2) only by a trivial increase in indices by one and hence also has value d_{FD} . Hence, we have shown that for m even, given any code with parameters m, N and K, we can construct a second code with parameters $\frac{m}{2}$, 2N, and 2K having the same minimum distance and thus the lemma is proved.

The last preliminary result which we shall need is the self-evident:

Lemma 4: $d(m, N, K) \leq d(m+1, N, K)$.

We are now in a position to prove the main result of this section.

Theorem 1: $d(m, N, K) \leq [\frac{m+5}{2}](N-K) + 1$, where the square brackets denote the integer part of the enclosed number.

Proof: Consider first the case when $(N-K)$ is even. Then

$$d(m, N, K) \leq d(m, N-K+1, 1) \leq (N-K+1) + \frac{m}{2}(N-K)$$

where the first inequality follows from lemma 1 and the second from lemma 2. Hence the inequality in the theorem actually holds with strict inequality in this case.

Next consider the case when $(N-K)$ is odd. Then

$$d(m, N, K) \leq d(m, N-K+1, 1) \leq d([\tfrac{m+1}{2}], 2N-2K+2, 2)$$

where the first inequality follows from lemma 1 and the second from the combination of lemmas 3 and 4. Again applying lemmas 1 and 2 in order to the last member of the preceding inequality, we obtain

$$d(m, N, K) \leq d([\tfrac{m+1}{2}], 2N-2K+1, 1) \leq 2(N-K) + 1 + [\tfrac{m+1}{2}](N-K)$$

which is equivalent to the inequality in the theorem. Thus the theorem is proved for all cases.

Recalling that $n_{FD} = (m+1)N$ and that $R = K/N$, we obtain immediately from Theorem 1:

Corollary 1: $$\lim_{m \to \infty} \frac{d(m, N, K)}{n_{FD}} \leq \frac{1}{2}(1-R) \quad .$$

Corollary 1, which is the asymptotic case of Theorem 1 for large constraint lengths n_{FD}, provides an upper bound for the ratio d_{FD}/n_{FD} that coincides with the usual asymptotic Plotkin upper bound [3] for the d_{min}/n ratio for a block code with rate R and constraint length n .

The key idea in deriving the preceding bound, namely the content of lemma 2, was first pointed out to the author by Jones [4] in 1962. The content of this lemma has also been independently stated by Lin and Lyne [5]. The remainder of the derivation, i.e. the necessary tricks to reduce the general case so that lemma 2 may be applied, was supplied by the author.

Wilder [6] has recently generalized the preceding derivations to convolutional codes defined over an arbitrary finite field $GF(q)$. In this general case, the righthand side of the inequality in corollary 1 becomes $\frac{q-1}{q}(1-R)$ which coincides with the asymptotic Plotkin bound for block codes. More surprisingly, in the binary case, Wilder found that the inequality in lemma 2 is an equality in many instances. For $N = 3$, Bussgang's tabulation [7] of optimal codes shows that equality is obtained for $m \leq 6$. Wilder found equality for $m \leq 4$

when $N = 5$, and equality for $m \leq 3$ for general N. Wilder also succeeded in generalizing lemma 2 to apply to a class of nonlinear tree codes of which convolutional codes are the linear special case.

3. Gilbert Lower Bound on Definite-Decoding Minimum Distance

A. The Gilbert Bound on Feedback-Decoding Minimum Distance

We digress momentarily to consider the well-known Gilbert lower bound on d_{FD}. For clarity, we treat only the case where $N = K + 1$, i.e. where $\underline{p}_u$ is a single binary digit p_u and $R = \frac{K}{K+1}$. In this case, the matrices G_j, which have dimension $(N - K) \times K$ in general, are simply K-dimensional row vectors. We shall emphasize this fact by writing the matrix G_j as $\underline{G}_j$. We shall here and hereafter use a prime to denote the transpose of a vector so that $\underline{G}_j'$ for instance is a K-dimensional column vector. With this notation, it then follows from (1) that

$$\begin{bmatrix} p_0 \\ p_1 \\ \vdots \\ p_m \end{bmatrix} = \begin{bmatrix} \underline{i}_0' & \underline{0}' & \cdots & \underline{0}' \\ \underline{i}_1' & \underline{i}_0' & \cdots & \underline{0}' \\ & \vdots & & \\ \underline{i}_m' & \underline{i}_{m-1}' & \cdots & \underline{i}_0' \end{bmatrix} \begin{bmatrix} \underline{G}_0' \\ \underline{G}_1' \\ \\ \underline{G}_m' \end{bmatrix} \tag{3}$$

and we shall refer to the matrix of information vectors in (3) as the $\underline{i}$-matrix. Note that the $\underline{i}$-matrix is an $(m+1) \times (m+1)K$ dimensional matrix of binary digits. We shall refer to the $(m+1)$-dimensional vector on the lefthand side of (3) as the p-vector, and shall refer to the $(m+1)K$-dimensional vector $(\underline{i}_0, \underline{i}_1, \ldots, \underline{i}_m)$ as the $\underline{i}$-vector.

A particular code will have $d_{FD} < d$ if and only if there is an $\underline{i}$-vector with $\underline{i}_0 \neq \underline{0}$ and a p-vector satisfying (3) whose combined Hamming weight is less than d. But since there are only $n_{FD} = (K+1)(m+1)$ positions in the combined vectors, there are only

$$\sum_{j=0}^{d-1} \binom{n_{FD}}{j} < 2^{n_{FD} H\left(\frac{d}{n_{FD}}\right)} \tag{4}$$

possible choices of low weight combinations. In obtaining (4), use has been made of the well-known inequality [8]

$$\sum_{j=0}^{\delta n} \binom{n}{j} \leq 2^{nH(\delta)} \quad \text{when} \quad \delta \leq \frac{1}{2} \tag{5}$$

where $H(\delta) = -\delta \log_2 \delta - (1-\delta) \log_2 (1-\delta)$ is the binary entropy function. Note next that $\underline{i}_0 \neq \underline{0}$ guarantees that the $\underline{i}$-matrix in (3) has rank exactly $m+1$ so that any combination of an $\underline{i}$-vector with $\underline{i}_0 \neq \underline{0}$ and a p-vector is a solution of (3) for a fraction exactly $2^{-(m+1)}$ of all the codes. Hence, it follows from (4) that if

$$2^{n_{FD} H\left(\frac{d}{n_{FD}}\right) - (m+1)} < 1 ,$$

then not all the codes have a combined $\underline{i}$-vector and p-vector with $\underline{i}_0 \neq \underline{0}$ with Hamming weight less than d . Hence it follows that there must exist at least one code such that its minimum distance satisfies

$$2^{n_{FD} H(d_{FD}/n_{FD}) - (m+1)} \geq 1 = 2^0 .$$

Since $(m+1) = n_{FD}(1-R)$, this inequality can be written

$$H(d_{FD}/n_{FD}) \geq 1 - R \tag{6}$$

for at least one convolutional code of rate R and constraint length n_{FD} . Inequality (6) is the usual asymptotic Gilbert bound [9] which holds for arbitrary $R = K/N$ although the derivation here has been restricted to $N = K + 1$.

B. The Gilbert Bound on d_{DD}

A second decoding method for convolutional codes, called <u>definite decoding</u> (DD) by Robinson [2], calls for the decoding estimate of $\underline{i}_u$ to be made without employing previous decoding estimates. The purpose of DD is to avoid the error-propagation effect inherent in feedback decoding. In particular, we assume that the decoding estimate of $\underline{i}_u$ is to be based on the received digits corresponding to $\underline{i}_{u-m}, \underline{i}_{u-m+1}, \ldots, \underline{i}_{u+m}$ and to $\underline{p}_u, \ldots, \underline{p}_{u+m}$, i.e. corresponding to the parity digits affected by $\underline{i}_u$ and to all the information digits affecting these parity digits. The number of these digits is the <u>definite-decoding constraint length</u>, n_{DD} .

$$n_{DD} = (2m+1)K + (m+1)(N-K) .$$

In order to make $u = 0$ typical of the general case, we abrogate our previous assumption that $\underline{i}_u = \underline{0}$ for $u < 0$ and hereafter allow these past information digits to assume arbitrary values. The <u>definite-decoding minimum distance</u>, d_{DD}, is then appropriately defined as the fewest number of positions that two encoded sequences with differing values of $\underline{i}_0$ are found to disagree over the DD constraint length. By the usual linearity argument, it then follows that

$$d_{DD} = \min_{\underline{i}_0 \neq \underline{0}} W_H(\underline{i}_{-m}, \underline{i}_{-m+1}, \ldots, \underline{i}_m, \underline{p}_0, \underline{p}_1, \ldots, \underline{p}_m) . \tag{7}$$

Comparison of (2) and (7) reveals that for any code $d_{DD} \le d_{FD}$. Hence, upper bounds on d_{FD} are a fortiori upper bounds on d_{DD}, but lower bounds on d_{FD} cannot be presumed to be lower bounds on d_{DD}.

Until further notice, we consider only the case $N = K + 1$ as was done in the preceding subsection. For this case, we have

$$n_{DD} = (2m + 1)K + (m + 1) \tag{8}$$

and from (3) as modified to account for the fact that we no longer assume $\underline{i}_u = \underline{0}$ for $u < 0$

$$\begin{bmatrix} p_0 \\ p_1 \\ \vdots \\ p_m \end{bmatrix} = \begin{bmatrix} \underline{i}'_0 & \underline{i}'_{-1} & \cdots & \underline{i}'_{-m} \\ \underline{i}'_1 & \underline{i}'_0 & \cdots & \underline{i}'_{-m+1} \\ & & \vdots & \\ \underline{i}'_m & \underline{i}'_{m-1} & \cdots & \underline{i}'_0 \end{bmatrix} \begin{bmatrix} \underline{G}'_0 \\ \underline{G}'_1 \\ \vdots \\ \underline{G}'_m \end{bmatrix} . \tag{9}$$

In the remainder of this subsection, we prove a lower bound on d_{DD} which we call a "Gilbert bound" not because of a formal similarity to (6) but because the method of proof will be along the same lines that led to (6). That some modification in the proof will be required is clear from the fact that when $\underline{i}_u = (1, 1, \ldots, 1)$, all u, then the $\underline{i}$-matrix in (9) has rank only 1 and hence fully 2^{-1} or one-half of all the codes have the p-vector $(0, 0, \ldots, 0)$ occuring in combination with this one particular $\underline{i}$-matrix. Fortunately, the combined p-vector and $\underline{i}$-vector [where we now take the $\underline{i}$-vector to be $(\underline{i}_{-m}, \underline{i}_{-m+1}, \ldots, \underline{i}_m)$] have high Hamming weight so that it does not follow that this one-half of the codes have small definite-decoding minimum distance. For ease of reference, the combined $\underline{i}$-vector and p-vector, i.e. the vector

$$(\underline{i}_{-m}, \underline{i}_{-m+1}, \ldots, \underline{i}_m, p_0, p_1, \ldots, p_m) ,$$

will be called the code-vector.

The argument that we shall use to obtain the Gilbert lower bound on d_{DD} runs roughly as follows: Let M_r be the number of code-vectors with $\underline{i}_0 \neq \underline{0}$ such that the $\underline{i}$-matrix has rank r and W_H(code-vector) $< d$. Then (assuming that it appears in some code) each such code vector appears in a fraction 2^{-r} of all the codes. Hence if

$$\sum_{r=1}^{m+1} M_r\, 2^{-r} < 1 \ ,$$

then there must exist at least one code such that $d_{DD} \geq d$.

1. Periodic i-Matrices and Linear FSR's

Definition 2: A periodic matrix is a matrix

$$\begin{bmatrix} \underline{a}'_0 & \underline{a}'_{-1} & \cdots & \underline{a}'_{-n} \\ \underline{a}'_1 & \underline{a}'_0 & \cdots & \underline{a}'_{-n+1} \\ \vdots & & & \\ \underline{a}'_m & \underline{a}'_{m-1} & \cdots & \underline{a}'_{m-n} \end{bmatrix} \qquad \begin{matrix} n \geq n \geq r \\ 1 \leq r \leq m \end{matrix} \qquad (10)$$

(where each $\underline{a}_j$ is a K-dimensional binary column vector) such that the matrix has rank r, its first r rows are linearly independent, and the linear combination of the first r rows which produces the $(r+1)$-st row includes the first row with a multiplier of 1 .

Thus if A_j denotes the $(j+1)$-st row of a periodic matrix (10), then

$$A_j = \sum_{g=1}^{r} c_g\, A_{j-g} \qquad (c_r = 1) \qquad j = r,\ r+1,\ \ldots,\ m\ . \qquad (11)$$

(11) holds by the definition of a periodic matrix when $j = r$; the form of (10) and the fact that A_j, $j > r$, must be linearly dependent on preceding rows guarantees that (11) must hold for $j > r$ as well. Conversely, if $r > 0$ is the smallest integer such that a recursion of the form (11) holds, then a matrix of the form (10) is periodic with rank r .

We see directly from (10) that (11) is fully equivalent to

$$\underline{a}_j = \sum_{g=1}^{r} c_g\, \underline{a}_{j-g} \qquad (c_r = 1) \qquad j = r-n,\ r-n+1,\ \ldots,\ m\ . \qquad (12)$$

Letting a_{jK+h-1}, $h = 1, 2, \ldots, K$, denote the h-th digit in $\underline{a}_j$, we see that (12) in turn is fully equivalent to

$$a_j = \sum_{g=1}^{r} c_g\, a_{j-gK} \quad (c_r = 1) \quad j = (r-n)K,\ (r-n)K+1,\ \ldots,\ mK+K-1\ .$$

(13)

We define the outer-fringe of a periodic matrix (10) to be the vector

$$(\underline{a}_{-n}, \underline{a}_{-n+1}, \ldots, \underline{a}_{m}) = (a_{-nK}, a_{-nK+1}, \ldots, a_{mK+K-1})$$

and note that this is an $(m+n+1)K$-dimensional column vector. The recursion (13) is just the statement that the outer-fringe is an $(m+n+1)$ digit output segment of a linear feedback shift-register (FSR) with tap connections every K-th stage as determined by $c_1, c_2, \ldots, c_r$. This FSR is shown in Figure 1. Note that since $c_r = 1$, the last stage of this rK-stage linear FSR is always tapped, i.e. the FSR is non-singular, and it is well-known in this case that all output sequences are periodic. This is the motivation for the nomenclature in definition 2, although of course the outer-fringe may not contain a complete period of an output sequence since the latter can in fact be as great as $K(2^r - 1)$. We note further that the outer-fringe cannot be an output segment of any such FSR with fewer than rK stages since in the latter case the periodic matrix (10) would be found to have rank less than r. We state the essential facts brought out in this discussion as:

Theorem 2: The outer-fringe of a rank r periodic matrix (10) is an $(m+n+1)K > 2rK$ digit output segment of a unique rK-stage nonsingular linear FSR and of no shorter linear FSR tapped only every K-th stage.

We next turn our attention to proving several facts about the output sequences of FSR's that will be exploited in the sequel. It will prove convenient to state these results in terms of the fractional weight of a vector $\underline{v}$ which we define to be the quantity $\frac{1}{n} W_H(\underline{v})$ where n is the dimension of $\underline{v}$.

Lemma 5: For any $n \geq L > 0$, and any δ, $0 < \delta \leq \frac{1}{2}$, the number of binary n-digit segments in any set such that each segment has fractional weight δ or less and no two segments coincide in any span of L consecutive digits is less than $2^{3LH(\delta)}$.

Proof: Let M be the maximum number of segments in such a set and suppose first that $L \leq n < 3L$. Since all segments must be distinct

$$M \leq \sum_{j=0}^{[\delta n]} \binom{n}{j} < \sum_{j=0}^{[3\delta L]} \binom{3L}{j} \leq 2^{3LH(\delta)}$$

where we have made use of (5) and [ ] throughout this proof denotes the integer part of the enclosed number. It remains to consider $n \geq 3L$. Let $n = iL + n'$ where i is the quotient and n' the remainder when n is divided by L. We note that no segment can have weight more than $[\delta n]$ and hence each segment must have weight $[\frac{4}{3}\delta iL]$ or less in its first iL positions. Suppose that

$$M \geq \sum_{j=0}^{m} \binom{L}{j} = M' \quad \text{where} \quad m = [\tfrac{8}{3}\delta L + 1] \ .$$

In each of the first i spans of L digits, the average weight of the M segments cannot be less than that of the M' distinct lowest weight vectors of length L, i.e. the $\binom{L}{j}$ vectors of weight j for $j = 0, 1, \ldots, m$. But, for any $k < \frac{1}{2}m$, the $\binom{L}{k}$ vectors of weight k are outnumbered by the $\binom{L}{m-k}$ vectors of weight $m-k$ so that the average weight of these M' vectors exceeds $\frac{1}{2}m$. Hence, in their first iL digits, the M segments have average weight exceeding $\frac{i}{2}m = \frac{i}{2}[\frac{8}{3}\delta L + 1] \geq [\frac{4}{3}\delta iL]$ contradicting the fact none has weight more than $[\frac{4}{3}\delta iL]$ over this span of iL digits. Hence we conclude that

$$M < \sum_{j=0}^{[\frac{8}{3}\delta L+1]} \binom{L}{j} \leq 2^{LH\left(\frac{[\frac{8}{3}\delta L+1]}{L}\right)} \leq 2^{LH(3\delta)} \leq 2^{3LH(\delta)}$$

where the third inequality holds under the further proviso that $\delta L \geq \frac{2}{3}$ since $[\frac{8}{3}x + 1] \leq 3x$ for any $x \geq \frac{2}{3}$. The remaining case where $\delta < \frac{2}{3}\frac{1}{L}$ is trivial and it is readily checked that the lemma is true in this case.

An immediate application of lemma 5 yields:

Lemma 6: For any $n \geq L$, and any δ, $0 < \delta \leq \frac{1}{2}$, of the 2^L distinct output segments of length n obtainable from an L-stage nonsingular linear FSR, fewer than $2^{3LH(\delta)}$ have fractional weight δ or less.

Proof: We simply note that any L consecutive digits in an output segment determine a state of the FSR (see Figure 1 which is the special case of an $L = rK$ stage register) so that any two segments which agree in such a span must agree everywhere thereafter. But since the output sequences of the FSR are periodic, the segments must also agree in their previous digits and hence must be the same segment. The lemma now follows directly from lemma 5.

Lemma 7: Given fixed values of m, n, K and r in definition 2, the number of distinct outer-fringes of rank r periodic matrices such that the outer-fringe has fractional weight δ or less, $0 < \delta \leq \frac{1}{2}$, is less than $2^{6KrH(\delta)}$.

Proof: It can be shown [10] that if the shortest linear FSR which can generate an n-digit, $n \geq 2L$, segment has length L, then any 2L successive digits in the segment uniquely determine the FSR. Hence, from theorem 2, we conclude that any 2Kr successive digits in the outer-fringe uniquely determine the entire outer-fringe. Thus there can be no more valid outer-fringes of fractional weight δ or less than there are $(m+n+1)K > 2rK$ digit segments of fractional weight δ or less such that no two coincide in any $2rK$ consecutive positions. By lemma 5, this number is less than $2^{3(2rK)H(\delta)} = 2^{6rKH(\delta)}$

We are now prepared to connect the notion of a periodic matrix with the $\underline{i}$ matrix in (9).

Theorem 3: If the $\underline{i}$-matrix in (9) has rank $r < \frac{m}{3}$, then the reduced $\underline{i}$-matrix

$$\begin{bmatrix} \underline{i}'_r & \underline{i}'_{r-1} & \cdots & \underline{i}'_{r-m} \\ \underline{i}'_{r+1} & \underline{i}'_r & \cdots & \underline{i}'_{r-m+1} \\ \vdots & & & \\ \underline{i}'_{m-r} & \underline{i}'_{m-r-1} & \cdots & \underline{i}'_{-r} \end{bmatrix} \tag{14}$$

is a periodic matrix of rank L, $L \leq r$, whenever $\underline{i}_0 \neq \underline{0}$.

Proof: Let I_j denote the $(j+1)$-st row in the i-matrix of (9) and let s be the least index such that I_s is a linear combination of preceding rows. Let I_{s-L} be the first row appearing with multiplier 1 in the unique combination of the first s rows which forms I_s, then

$$I_s = \sum_{g=1}^{L} c_g I_{s-g} \quad (c_L = 1) \tag{15}$$

and we note that $L \leq s \leq r$. If $r = s$, all rows after I_s must also satisfy the recursion (15) and hence the theorem is trivially true. If not, suppose that $t > s$ is the least index such that (15) is not satisfied, i.e.

$$I_j = \sum_{g=1}^{L} c_g I_{j-g} \quad (c_L = 1) \quad j = s, s+1, \ldots, t-1 \tag{16}$$

but

$$I_t \neq \sum_{g=1}^{L} c_g I_{t-g} \quad (c_L = 1) \;. \tag{17}$$

In this case, we claim that the $\underline{i}$-matrix in (9) has rank exactly $(m+1) + (t-s)$. To show this, note that (16) and (17) are equivalent to

$$\underline{i}_j = \sum_{g=1}^{L} c_g \underline{i}_{j-g} \qquad j = s-m, s-m+1, \ldots, t-1 \tag{18}$$

and

$$\underline{i}_t \neq \sum_{g=1}^{L} c_g \underline{i}_{t-g} \ . \tag{19}$$

Now suppose that I_u, for any $u \geq t$, can be written as a linear combination of preceding rows, i.e.

$$I_u = \sum_{h=1}^{u} a_h I_{u-h} \ . \tag{20}$$

But (20) is equivalent to

$$\underline{i}_j = \sum_{h=1}^{u} a_h \underline{i}_{j-h} \qquad j = u-m, u-m+1, \ldots, u \tag{20'}$$

which in particular, since $u-m < t \leq u$, implies

$$\underline{i}_t = \sum_{h=1}^{u} a_h \underline{i}_{t-h} \ . \tag{21}$$

But the terms in the summation on the righthand side of (21) involve only $\underline{i}_j$ for j in the range such that (18) is valid. Hence we may use (18) in (21) to obtain

$$\underline{i}_t = \sum_{h=1}^{u} a_h \sum_{g=1}^{L} c_g \underline{i}_{t-h-g} = \sum_{g=1}^{L} c_g \sum_{h=1}^{u} a_h \underline{i}_{t-g-h} \ . \tag{22}$$

We now recognize, since $t-u-L > u-m$, that (20') may be used to rewrite the righthand side of (22) which yields

$$\underline{i}_t = \sum_{g=1}^{L} c_g \underline{i}_{t-g} \tag{23}$$

and hence gives a contradiction of (19). We conclude that the only rows in the $\underline{i}$-matrix (9) which can be written as linear combinations of preceding rows are the $t-s$ rows satisfying (16). Since the matrix has $m+1$ rows, its rank then is exacly $(m+1) - (t-s)$ as claimed. But the rank r is given as less than $\frac{m}{3}$. We have already noted that $s \leq r$, and hence $t = (m+1) - r + s > m - r$. Hence all the rows in the reduced $\underline{i}$-matrix (14) are rows which satisfy the recursion (16) and hence this $\underline{i}$-matrix is periodic as claimed. Moreover it has rank $L \leq r$.

From (9), we find that

$$\begin{bmatrix} p_r \\ p_{r+1} \\ \vdots \\ p_{m-r} \end{bmatrix} = \begin{bmatrix} \underline{i}'_r & \underline{i}'_{r-1} & \cdots & \underline{i}'_{r-m} \\ \underline{i}'_{r+1} & \underline{i}'_r & \cdots & \underline{i}'_{r-m+1} \\ & & \vdots & \\ \underline{i}'_{m-r} & \underline{i}'_{m-r-1} & \cdots & \underline{i}'_{-r} \end{bmatrix} \begin{bmatrix} \underline{G}'_0 \\ \underline{G}'_1 \\ \vdots \\ \underline{G}'_m \end{bmatrix} . \tag{24}$$

We call the lefthand side of (24) the <u>reduced</u> <u>p-vector</u> and note that it is $m - 2r + 1$ component vector uniquely determined by the reduced $\underline{i}$-matrix. The outer-fringe of the reduced $\underline{i}$-matrix is a $2(m-r)K+K$ component vector that we call the <u>reduced $\underline{i}$-vector</u>. The combination of the reduced p-vector and reduced $\underline{i}$-vector will be called the <u>reduced code-vector</u>.

<u>Lemma 8</u>: If the reduced $\underline{i}$-matrix is periodic of rank L, then the reduced p-vector is an output segment of an L-stage nonsingular linear FSR uniquely determined by the reduced $\underline{i}$-matrix. In particular,

$$p_j = \sum_{g=1}^{L} c_g p_{j-g} \quad (c_L = 1) \qquad j = r+L, r+L+L, \ldots, m-r \tag{25}$$

where c_g, $g = 1, 2, \ldots, L$, are the FSR connections uniquely determined by the reduced $\underline{i}$-matrix.

<u>Proof</u>: From (13) we see that the digits in each <u>column</u> of the reduced $\underline{i}$-matrix satisfy the recursion (25). But (24) shows that the reduced p-vector is always a linear combination of these columns and hence also satisfies the recursion (25).

We are finally in a position to tie all the preceding results together so as to obtain a Gilbert bound on d_{DD}.

We begin by noting that for $r \leq \Delta(m+1)$, where Δ, $0 < \Delta < \frac{1}{3}$, will be chosen later, if the reduced p-vector has fractional weight δ_p, then the entire code-vector has fractional weight δ' satisfying

$$\delta' \geq \frac{m-2r+1}{n_{DD}} \delta_p > \frac{1-2\Delta}{2K+1} \delta_p . \tag{26}$$

Similarly, if the reduced $\underline{i}$-vector has fractional weight δ_i, then

$$\delta' \geq \frac{2(m-r)K+K}{n_{DD}} \delta_i > (1-2\Delta)\frac{2K}{2K+1} \delta_i \qquad (27)$$

where the last inequality requires the proviso

$$m \geq \frac{1-2\Delta}{\Delta}$$

and henceforth we assume that we are considering only m sufficiently large to satisfy this inequality.

For a given δ, we wish to demonstrate the existence of a code such that $d_{DD} \geq \delta n_{DD}$. We begin by choosing

$$\delta_i = \frac{2K+1}{2K} \frac{1}{1-2\Delta} \delta < \frac{1}{2} \qquad (28)$$

and

$$\delta_p = \frac{2K+1}{1-2\Delta} \delta < \frac{1}{2} . \qquad (29)$$

We next divide the set of all possible code-vectors having $\underline{i}_0 \neq \underline{0}$ and fractional weight δ or less into two sets S_1 and S_2 defined as follows. S_1 contains only those code-vectors such that the $\underline{i}$-matrix has rank r satisfying $r \geq \Delta(m+1)$ and S_2 contains those for which the $\underline{i}$-matrix has rank r, $r < \Delta(m+1)$.

First consider the set S_1. S_1 cannot contain more than all of the code vectors of fractional weight δ or less, and each vector in S_1 appears in a fraction at most $2^{-(m+1)\Delta}$ of all codes. Hence the fraction F_1 of codes which contain any vector in S_1 satisfies

$$F_1 \leq \sum_{j=0}^{[\delta n_{DD}]} \binom{n_{DD}}{j} 2^{-(m+1)\Delta} \leq 2^{-n_{DD}\{\frac{\Delta}{2K+1} - H(\delta)\}} \qquad (30)$$

where here and hereafter we use [ ] to indicate the integer part of the enclosed number.

The consideration of set S_2 becomes considerably more interesting. From (26) and (27) we conclude that any vector in S_2 must have <u>both</u> fractional weight δ_i or less in its reduced $\underline{i}$-vector <u>and</u> fractional weight δ_p or less in its reduced p-vector. Hence, the number of distinct reduced code-vectors found within the vectors in S_2 such that the reduced $\underline{i}$-matrix has some given rank L is less than

$$2^{6KLH(\delta_i)}\,2^{3LH(\delta_p)} = 2^{6KLH(\delta_i)+3LH(\delta_p)}$$

which follows from the facts that lemma 7 gives the first factor as bounding the number of reduced $\underline{i}$-vectors to be considered whereas lemmas 6 and 8 give the second factor as bounding the number of p-vectors to be considered with any given $\underline{i}$-vector. Note also that the reduced $\underline{i}$-vector is a non-zero output segment from a KL-stage non-singular linear FSR and hence must have at least one non-zero digit every KL digits which requires that it have fractional weight exceeding $\frac{1}{K2L}$. Thus S_2 contains no reduced $\underline{i}$-vector such that $L \le \frac{1}{2K\delta_i}$. But the fraction of codes containing any reduced code-vector such that the reduced $\underline{i}$-vector in (24) has rank L is at most 2^{-L}. We conclude then that the fraction F_2 of codes containing any code-vector in S_2 satisfies

$$F_2 < \sum_{L=[\frac{1}{2\delta_i K}+1]}^{\infty} 2^{6KLH(\delta_i)+3LH(\delta_p)-L} \quad . \tag{31}$$

With the aid of (28) and (29) and from the convexity of the entropy function, we obtain

$$F_2 < \sum_{L=[\frac{1-2\Delta}{2K+1}\frac{1}{\delta}+1]}^{\infty} 2^{-L\{1-6\frac{2K+1}{1-2\Delta}H(\delta)\}}$$

which upon summing of the geometric series yields

$$F_2 < 2^{-\frac{6}{\delta}\{\frac{1-2\Delta}{6(2K+1)}-H(\delta)\}} \tag{32}$$

provided that

$$H(\delta) < \frac{1}{6}\,\frac{1-2\Delta}{2K+1} \quad . \tag{33}$$

Combining (30) and (32), under the proviso of (33), we find that the fraction of codes containing any element of S_1 <u>or</u> S_2 is at most

$$F_1 + F_2 < 2^{-n_{DD}\{\frac{\Delta}{2K+1} - H(\delta)\}} + 2^{-\frac{6}{\delta}\{\frac{1-2\Delta}{6(2K+1)} - H(\delta)\}} . \quad (34)$$

We now exercise the free choice that we have reserved and choose $\Delta = \frac{1}{6}$. (34) then becomes

$$F_1 + F_2 < 2^{-n_{DD}\{\frac{1}{6(2K+1)} - H(\delta)\}} + 2^{-\frac{6}{\delta}\{\frac{1}{9(2K+1)} - H(\delta)\}} . \quad (35)$$

We next choose δ to satisfy

$$H(\delta) < \frac{1}{10}\frac{1}{2K+1} \quad (36)$$

which we note is consistent with our proviso (33) and is also sufficient to guarantee that the first term in (35) vanishes as n_{DD} gets large. It remains to show that the second term in (35) is less than 1 . To see this we note that this term has its maximum value when $K = 1$ and (36) holds with equality in which case the term can be evaluated to be

$$2^{-\frac{6}{.0035}\{\frac{1}{27} - \frac{1}{30}\}} < 2^{-6} .$$

Hence, whenever (36) is satisfied, not all codes contain code vectors with $\underline{i}_0 \neq \underline{0}$ and fractional weight $\delta\, n_{DD}$ or less. We conclude that there exists as least one code with definite-decoding minimum distance d_{DD} satisfying

$$H(\frac{d_{DD}}{n_{DD}}) \geq \frac{1}{10}\frac{1}{2K+1}$$

for every n_{DD} sufficiently large. We state this result as:

Theorem 4: For $N = K + 1$ (and hence $R = \frac{K}{K+1}$), and for all n_{DD} sufficiently large, there exists at least one convolutional code such that

$$H(\frac{d_{DD}}{n_{DD}}) \geq \frac{1}{10}\frac{1}{2K+1} = \frac{1}{10}\frac{1-R}{1+R} \quad (37)$$

Theorem 4 provides our long-sought Gilbert bound for the special case when $N = K + 1$. We now sketch the manner in which this bound can be extended to arbitrary $N > K$. In the general case, there are $N - K$ p-vectors, each of which, say the h-th, satisfies (9) with $\underline{G}_j$ interpreted as the h-th row of the matrix G_j, $h = 1, 2, \ldots, N - K$. Thus each of these reduced p-vectors is an output sequence of the <u>same</u> linear FSR and hence the number of distinct reduced code-vectors in S_2 is less than $2^{6KLH(\delta_i) + 3(N-K)LH(\delta_p)}$. Equations (28) and (29) are now replaced by $\delta_i = \frac{N+K}{2K}\frac{1}{1-2\Delta}\delta$ and $\delta_p = \frac{N+K}{N-K}\frac{1}{1-2\Delta}$. The other arguments go through virtually unchanged and result in:

<u>Theorem 5</u>: For any $N > K$ and for all n_{DD} sufficiently large, there exist convolutional codes of rate $R = \frac{K}{N}$ and definite-decoding minimum distance d_{DD} such that

$$H\left(\frac{d_{DD}}{n_{DD}}\right) \geq \frac{1}{10}\frac{1 - R}{1 + R} \quad . \tag{38}$$

4. Remarks

Robinson [2] has given what he calls a 'Gilbert bound' on d_{DD}, but this bound is not asymptotically useful since it shows d_{DD} growing less than linearly with n_{DD} for large n_{DD}. Wagner [11] has obtained an asymptotically useful bound on d_{DD} for a class of time-varying codes, viz. codes such that G_j in (1) is a periodic function of u . Wagner also used a different, and less natural, definition of the definite-decoding constraint length from that employed herein and his resultant bound was useful only for $R < \frac{1}{2}$. For Wagner's codes, but using our definition of constraint length, Morrisse and Costello [12] have recently obtained the bound

$$H\left(\frac{d_{DD}}{n_{DD}}\right) \geq \frac{1 - R}{1 + R} \tag{39}$$

for all n_{DD} sufficiently large. Time-varying codes are an artifice to avoid the rank problems encountered with ordinary convolutional codes that had to be surmounted in this paper, but appear to be of little practical interest due to the increased instrumentation complexity. We conjecture that tighter bounding arguments, particularly in our lemma 5 which is especially crude, can eventually do away with the additional factor of ten in (38) compared to (39). We could have improved this factor somewhat in the present instance, but only at the expense of complicating several proofs beyond what any reader could endure. We have settled for obtaining what we believe is the functional form of the tightest possible bound without concerning ourselves overly about the constant multiplier.

Lemmas 5 through 7 were proved some time ago by the author and used by Kolor [13] to obtain what we believe is the first asymtotically useful bound on d_{DD} for ordinary convolutional codes. Kolor restricted himself to the special case $K = 1$, $N = 2$. It is difficult to compare the $K = 1$ case in theorem 3 to Kolor's result since Kolor quite sensibly ignored the 'integer part' difficulties which to overcome rigorously caused our lemma 5 to be a very loose bound. Kolor's major result was a decomposition theorem for parasymmetric matrices (matrices of the form (10) with $K = 1$) which is essentially embodied in our theorem 3. In most respects, the material in section 3 is a generalization and simplification of the method used by Kolor for $K = 1$ and $N = 2$, as well as an attempt to put the theory in a rigorous framework.

Finally, it should be mentioned that Robinson [14] has proved an upper bound on d_{FD} that is asymptotically the same as the bound in section 2 and also reduces to the Plotkin block coding bound for $m = 0$. In the nonasymptotic case, for small $N - K$, the bound in section 2 is generally superior to Robinson's bound, but is inferior in general when $N - K$ is large. The virtue of the bound in section 2 is its conceptual simplicity and ease of derivation as compared to Robinson's bound.

REFERENCES

[1] A. D. Wyner, "On the Equivalence of Two Convolution Code Definitions," IEEE Trans. on Inf. Th., IT-11, pp. 600-602, October 1965.

[2] J. P. Robinson, "Error Propagation and Definite Decoding of Convolutional Codes," IEEE Trans. on Inf. Th., IT-14, pp. 121-128, January 1968.

[3] W. W. Peterson, Error-Correcting Codes, M. I. T. Press and Wiley, pp. 48-50, 1961.

[4] D. M. Jones, Private Communication, RCA Missile Electronics and Control Division, Burlington, Mass., 1962.

[5] S. Lin and H. Lyne, "Some Results on Binary Convolutional Code Generators," IEEE Trans. on Inf. Th., IT-13, pp. 134-139, January 1967.

[6] W. C. Wilder, "Extensions of the Plotkin Bound," Course VI S. M. Thesis, M. I. T., Cambridge, Mass., August 1967.

[7] J. J. Bussgang, "Some Properties of Binary Convolutional Code Generators," IEEE Trans. on Inf. Th., IT-11, pp. 90-100, January 1965.

[8] J. M. Wozencraft and B. Reiffen, Sequential Decoding, M. I. T. Press and Wiley, (see appendix), 1961.

[9] J. L. Massey, Threshold Decoding, M. I. T. Press, pp. 15-17, 1963.

[10] J. L. Massey, "Shift-Register Synthesis and BCH Decoding, to appear in IEEE Trans. on Inf. Th.

[11] T. J. Wagner, "A Gilbert Bound for Periodic Binary Convolutional Codes," to appear in IEEE Trans. on Inf. Th.

[12] T. Morissey and D. Costello, Private Communication, Univ. of Notre Dame, Notre Dame, Ind., 1968.

[13] R. W. Kolor, "A Gilbert Bound for Convolutional Codes," Course VI S. M. Thesis, M. I. T., Cambridge, Mass., August 1967.

[14] J. P. Robinson, "An Upper Bound to the Minimum Distance of a Convolutional Code," IEEE Trans. on Inf. Th., IT-11, pp. 567-571, October 1965.

This work was supported in part by the National Aeronautics and Space Administration (NASA Grant NGR 15-004-026 to the Univ. of Notre Dame) under liaison with the Flight Data Systems Branch of the Goddard Space Flight Center, and in part by the National Aeronautics and Space Administration (Grant NsG-334) and the Joint Services Electronics Program ((Contract DA208-043-AMC-02536)E) at the Research Laboratory of Electronics, Mass. Inst. of Tech.

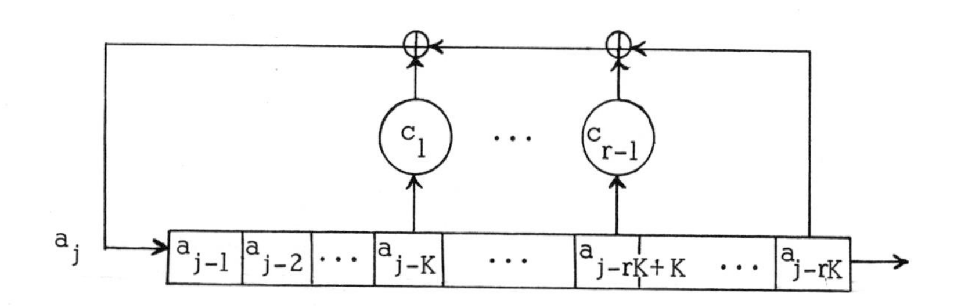

Figure 1. The rK-stage nonsingular linear feedback shift-register associated with a rank r periodic matrix.

CHARLES SALTZER

Topological Codes

1. Introduction

A general method for the construction of a class of linear codes, to be called topological codes, which are based on groups associated with a complex [1] is presented in this paper. A two-parameter subclass of these codes, called simplicial codes, is studied in detail. The information rates, the minimum weights and the word length are determined for simplicial codes. An important feature of the simplicial codes is that they can be completely orthogonalized [2], and hence, can be decoded by threshold decoding algorithms. The construction of the orthogonal parity-check equations for the simplicial codes is also given. Although the information rates for these codes are not high, the ease of decoding may make them interesting [3, 4]. Another noteworthy property of these codes is that codes with any prescribed minimum weights are easily constructed.

The graph-theoretic codes of D.A. Huffman [5], and of S.L. Hakimi and H. Frank [6] are special cases of the topological codes which are constructed from a one-dimensional complex.

The connection of these codes with the groups of algebraic topology is of interest in that it indicates a type of mathematical structure which gives some insight into the problem of decoding.

The paper is self-contained as far as the algebraic topology is concerned. The mathematical preliminaries are given in Section 2. The construction of the topological codes, and in particular, the simplicial codes is given in Section 3. The majority decoding and the minimum weight of simplicial codes is examined in Section 4. The Euler-Poincaré formula is established and the word length and information rates for topological codes and simplicial codes are determined in terms of Betti numbers in Sections 5 and 6. In the last section which is an appendix the Betti numbers are computed for a simplex.

2. Vector Spaces Associated with Simplicial Complexes

We start with a set of objects $x_0^0, x_1^0, \ldots,$ called points or vertices. An abstract p-simplex, where $p \geq 0$, is defined as a subset consisting of $(p+1)$ of these points which are called the vertices of the simplex. A p-simplex or simplex of dimension p is denoted by the symbol $(x^0_{i_0}, x^0_{i_1}, \ldots, x^0_{i_p})$. A non-empty subset of the vertices of a simplex is said to be a q-face of that simplex if it consists of $(q+1)$ vertices. A 1-simplex may be thought of as a line segment, a 2-simplex, as a triangle, and a 3-simplex, as a tetrahedron. An abstract complex K with a given set of vertices $x^0_0, x^0_1, \ldots, x^0_{r_0-1}$, where r_0 is the number of vertices in K, is a collection of simplexes which contains every face of each simplex of the collection. The p-simplexes of a complex are also denoted by $x^p_0, x^p_1, \ldots, x^p_{r_p-1}$ where r_p is the number of p-simplexes in the complex.

A p-chain y^p on a given complex is a function on the p-simplexes of the complex and is written as a linear form. Thus, the p-chain

$$y^p = c_0 x^p_0 + c_1 x^p_1 + \ldots + c_{r_p-1} x^p_{r_p-1}$$

assigns the value c_j to the simplex x^p_j. The value of the function y^p on the simplex x^p_j is called the coefficient of x^p_j in the chain y^p. If the coefficients are elements of a field, which is restricted here to the integers modulo 2, we can define the scalar multiple of a p-chain and the sum of p-chains in the following way. Let

$$y^p_1 = c_{10} x^p_0 + c_{11} x^p_1 + \ldots$$

$$y^p_2 = c_{20} x^p_0 + c_{21} x^p_1 + \ldots .$$

Then cy^p_1 and $y^p_1 + y^p_2$ are defined by

$$cy^p_1 = cc_{10} x^p_0 + cc_{11} x^p_1 + \ldots ,$$

$$y^p_1 + y^p_2 = (c_{10} + c_{20}) x^p_0 + (c_{11} + c_{21}) x^p_1 + \ldots .$$

In view of these definitions it follows that the p-chains constitute a vector space of dimension r_p which is denoted by V^p.

The boundary operator ρ is a linear operator which maps V^p into V^{p-1} and has the following value for a p-simplex:

$$\rho(x^0_{j_0} x^0_{j_1}, \ldots, x^0_{j_p}) = \sum_{q=0}^{p} (x^0_{j_0}, \ldots, x^0_{j_{q-1}}, x^0_{j_{q+1}}, \ldots, x^0_{j_p}), \quad (2.1)$$

.e. the sum of all the (p-1)-faces of the p-simplex. The space V^{-1} s taken as consisting of the single element 0 . The kernel of ρ, .e. the p-chains whose boundaries are 0 is denoted by Z^p and the hains in Z^p are called cycles. The image of V^p is denoted by ρV^p .

If the boundary operator is applied again in equation (2.1), the ight hand side will consist of all the (p-2)-faces, and each (p-2)-ace will occur exactly twice. Thus the boundary of the boundary of a -simplex is zero. By the linearity of ρ it follows that for any p-hain y^p ,

$$\rho\rho y^p = 0 \quad . \tag{2.2}$$

ince $\rho\rho V^{p+1} = 0$ we see that ρV^{p+1} is a subspace of Z^p . The ele-ıents of ρV^{p+1} are called bounding cycles.

The boundary operator is also described by the $p^{\underline{th}}$-incidence ıatrix

$$a^p_{ij} = \begin{cases} 1 & \text{if} \quad x^{p-1}_j \text{ is a face of } x^p_i \\ 0 & \text{if} \quad 0 \text{ if not} \end{cases} \tag{2.3}$$

$= 0, 1, \ldots, (r_p - 1)$; $j = 0, 1, \ldots, (r_{p-1} - 1)$. In terms of the elements f the incidence matrix we have

$$\rho x^p_i = \sum_{j=0}^{r_{p-1}-1} a^p_{ij} x^{p-1}_j , \qquad i = 0, 1, \ldots, (r_p - 1) \quad . \tag{2.4}$$

The coboundary operator δ is a linear operator which maps $^{p-1}$ into V^p and is defined by

$$\delta x^{p-1}_j = \sum_{i=0}^{r_{p-1}} a^p_{ij} x^p_i \quad . \tag{2.5}$$

'hus the coboundary of a (p-1)-simplex consists of the sum of those -simplexes whose boundaries contain the given (p-1)-simplex. If ı is the least upper bound of the dimensions of the simplexes of the iven complex then V^{m+1} is defined as the space consisting of 0 , nd the coboundary of every m-chain is taken as 0 .

If the coboundary of a p-chain is 0, the p-chain is said to e a p-cocycle. The set of p-cocycles is a vector subspace of V^p enoted by $\bar{Z}^p$. This is the kernel of δ restricted to V^p . If for p-chain y^p_1 there is a (p-1) chain y^{p-1}_2 such that $y^p_1 = \delta y^{p-1}_2$ we say that y_1^p is a p-coboundary. The set of p-coboundaries is a vector subspace of V^p and is denoted by δV^{p-1} .

If y_1^p and y_2^p are the p-chains

$$y_1^p = b_0 x_0^p + \ldots + b_{r_p-1} x_{r_p-1}^p$$

$$y_2^p = c_0 x_0^p + \ldots + c_{r_p-1} x_{r_p-1}^p ,$$

the scalar product of y_1^p and y_2^p is denoted by (y_1^p, y_2^p) and is defined by

$$(y_1^p, y_2^p) = \sum_{j=0}^{r_p-1} b_j c_j . \tag{2.6}$$

Since the arithmetic is arithmetic modulo 2 this product is not positive definite. The following rules are a direct consequence of the definition:

$$(y_1^p, y_2^p) = (y_2^p, y_1^p) \tag{2.7}$$

$$(cy_1^p, y_2) = c(y_1^p, y_2^p) \tag{2.8}$$

$$(y_1^p, y_2^p + y_3^p) = (y_1^p, y_2^p) + (y_1^p, y_3^p) . \tag{2.9}$$

If $(y_1^p, y_2^p) = 0$ we say that the p-chain y_1^p is orthogonal to the p chain y_2^p . Since $(y_1^p, x_j^p) = b_j$ we see that

$$y_1^p = \sum_{j=0}^{r_p-1} (y_1^p, x_j^p) x_j^p , \tag{2.10}$$

i.e. a p-chain is represented by its "Fourier series." Hence if a p-chain y^p is orthogonal to all p-chains it is 0 .

If we note that

$$(\rho x_i^p, x_k^{p-1}) = (\sum_{j=0}^{r_{p-1}-1} a_{ij}^p x_j^{p-1}, x_k^{p-1}) = a_{ik}^p ,$$

$$(x_i^p, \delta x_k^{p-1}) = (x_i^p, \sum_{j=0}^{r_p-1} a_{jk}^p x_j^p) = a_{ik}^p ,$$

it follows from (2.8) and (2.9) that for any p-chain y_1^p and any (p-1)-chain y_2^{p-1} we have

$$(\rho y_1^p, y_2^{p-1}) = (y_1^p, \delta y_2^{p-1}) . \tag{2.11}$$

Now for every (p+1)-chain y_1^{p+1},

$$(y_1^{p+1}, \delta\delta y_2^{p-1}) = (\rho y_1^{p+1}, \delta y_2^{p-1}) = (\rho\rho y_1^{p+1}, y_2^{p-1}) = 0 ,$$

since $\rho\rho y_1^{p+1} = 0$.
Hence

$$\delta\delta y_2^{p-1} = 0 . \tag{2.12}$$

This shows that a p-coboundary is a p-cocycle. Thus the p-coboundary space δV^{p-1} is a subspace of the p-cocycle space $\bar{Z}^p$.

If U and V are subspaces of a finite-dimensional vector space W, in which orthogonality is defined, we say that W is the quasi-direct sum of U and V,

$$W \sim U \oplus V ,$$

provided every vector of U is orthogonal to every vector of V and

$$\text{dimension } W = \text{dimension } U + \text{dimension } V .$$

It should be noted that for the spaces which occur in this paper U may be a subspace of V . If three subspaces of a vector space are orthogonal in pairs and the sum of their dimensions is the dimension of the given space we say that the given space is the quasi-direct sum of the three subspaces.

If U is the set of all vectors orthogonal to a subspace V of W then W is the quasi-direct sum of U and V if the inner product is defined as in equation (2.6). We shall call U the quasi-orthogonal complement of V with respect to W .

We show next that δV^{p-1} is the quasi-orthogonal complement of Z^p in V^p, and that ρV^{p+1} is the quasi-orthogonal complement of $\bar{Z}^p$ in V^p . From this it would follow that V^p is the quasi-direct sum of the subspaces Z^p and δV^{p-1}, i.e.

$$V^p \sim Z^p \oplus \delta V^{p-1} , \tag{2.13}$$

and also,

$$V^p \sim \bar{Z}^p \oplus \rho V^{p+1} . \tag{2.14}$$

To prove that Z^p and δV^{p-1} are quasi-orthogonal complements, let y_1^p be a p-cycle. If y_2^p is a p-coboundary then there is at least one (p-1)-chain y_3^{p-1} such that $y_2^p = \delta y_3^{p-1}$. Since y_1^p is a p-cycle we have by (2.11)

$$(y_1^p, \delta y_3^{p-1}) = (\rho y_1^p, y_3^{p-1}) = 0 .$$

This shows that a p-cycle is orthogonal to all p-coboundaries. Conversely, if y_1^p is orthogonal to all p-coboundaries, i.e., for every (p-1)-chain y_4^{p-1} we have

$$(y_1^p, \delta y_4^{p-1}) = 0$$

then by (2.11),

$$(\rho y_1^p, y_4^{p-1}) = 0 .$$

Since this is valid for every (p-1)-chain, the rule following equation (2.10) permits us to conclude that $\rho y_1^p = 0$, i.e., y_1^p is a p-cycle. The proof of the assertion that $\overline{Z}^p$ and ρV^{p+1} are quasi-orthogonal complements is similar.

3. The Construction of the Simplicial Codes

In the general case for a complex K and for each integer p, $0 < p < m$, we have four topological codes which are subspaces of V^p. The p-cycle code consists of all the p-cycles of Z^p. The word length n is the dimension of V^p which equals the number r_p of p-simplexes in the complex K. The number k of information symbols is the dimension of the p-cycle space Z^p. The p-coboundary space δV^{p-1} is the dual code of the p-cycle code. The p-cocycle space $\overline{Z}^p$ and the p-boundary space ρV^{p+1} define the two dual codes, the p-cocycle code and the p-cycle code. A code will be said to be a simplicial code if the complex K consists of a single simplex together with all of its faces.

Although much of the theory below is valid for general complexes we shall restrict ourselves to simplicial codes. Since for $p > 0$, every p-cycle of a simplex is a bounding cycle we have

$$Z^p = \rho V^{p+1} \qquad (p = 1, 2, \ldots, m-1) \qquad (3.1)$$

$$\overline{Z}^p = \delta V^{p-1} \qquad (p = 1, 2, \ldots, m-1) . \qquad (3.2)$$

Thus, in the case where the complex K is a simplex together with its faces, we get only two dual codes for each p.

As an illustration, consider the complex consisting of a three dimensional simplex together with its faces. The vertices are x_0, x_1, x_2, x_3 . Since every subset of these vertices is a face we have

$$r_0 = \binom{4}{1} = 4$$

$$r_1 = \binom{4}{2} = 6$$

$$r_2 = \binom{4}{3} = 4 \ .$$

We may designate the six 1-faces in the following way:

$$\begin{aligned} x_0^1 &= (x_0^0, x_3^0),\ x_1^1 = (x_1^0, x_3^0),\ x_2^1 = (x_2^0, x_3^0) \\ x_3^1 &= (x_0^0, x_1^0),\ x_4^1 = (x_1^0, x_2^0),\ x_5^1 = (x_0^0, x_2^0) \ . \end{aligned} \tag{3.3}$$

The four two faces are denoted by

$$\begin{aligned} x_0^2 &= (x_0^0, x_1^0, x_3^0),\ x_1^2 = (x_1^0, x_2^0, x_3^0) \\ x_2^2 &= (x_0^0, x_1^0, x_2^0),\ x_3^2 = (x_0^0, x_2^0, x_3^0) \ . \end{aligned} \tag{3.4}$$

To illustrate the computation of the incidence matrices consider the case $p = 2$. Now

$$\rho x_0^2 = \rho(x_0^0, x_1^0, x_3^0) = (x_0^0, x_1^0) + (x_0^0, x_3^0) + (x_1^0, x_3^0) = x_3^1 + x_0^1 + x_1^1 \ . \tag{3.5}$$

Similarly

$$\rho x_1^2 = x_1^1 + x_2^1 + x_4^1,\quad \rho x_2^2 = x_3^1 + x_4^1 + x_5^1,\quad \rho x_3^2 = x_0^1 + x_2^1 + x_5^1 \ . \tag{3.6}$$

Thus the 2-incidence matrix is

$$\begin{pmatrix} 1 & 1 & 0 & 1 & 0 & 0 \\ 0 & 1 & 1 & 0 & 1 & 0 \\ 0 & 0 & 0 & 1 & 1 & 1 \\ 1 & 0 & 1 & 0 & 0 & 1 \end{pmatrix} \ .$$

The coboundary operator for 1-simplexes can be read off from the columns of this matrix:

$$
\begin{aligned}
\delta x^1_0 &= x^2_0 + x^2_3 \\
\delta x^1_1 &= x^2_0 + x^2_1 \\
\delta x^1_2 &= x^2_1 + x^2_3 \\
\delta x^1_3 &= x^2_0 + x^2_2 \\
\delta x^1_4 &= x^2_1 + x^2_2 \\
\delta x^1_5 &= x^2_2 + x^2_3
\end{aligned}
\tag{3.7}
$$

If we represent a vector by its coefficients, for example,

$$\rho x^2_0 = x^1_0 + x^1_1 + 0x^1_2 + x^1_3 + 0x^1_4 + 0x^1_5 = 110100 \tag{3.8}$$

we can write the generating matrix for the 1-cycle code in the usual manner.

$$\begin{pmatrix} 1 & 1 & 0 & 1 & 0 & 0 \\ 0 & 1 & 1 & 0 & 1 & 0 \\ 0 & 0 & 0 & 1 & 1 & 1 \end{pmatrix} .$$

The 1-incidence matrix is obtained in a similar way, e.g.,

$$\rho x^1_0 = x^0_0 + x^0_3 = 1001 \quad . \tag{3.9}$$

Thus the 1-incidence matrix is

$$\begin{pmatrix} 1 & 0 & 0 & 1 \\ 0 & 1 & 0 & 1 \\ 0 & 0 & 1 & 1 \\ 1 & 1 & 0 & 0 \\ 0 & 1 & 1 & 0 \\ 1 & 0 & 1 & 0 \end{pmatrix} .$$

The coboundary operator for the vertices is obtained from the columns of this matrix, e.g.,

$$\begin{aligned} \delta x_0^0 &= x_0^1 + x_3^1 + x_5^1 = 100101 \\ \delta x_1^0 &= 010110 \\ \delta x_2^0 &= 001011 \\ \delta x_3^0 &= 111000 \ . \end{aligned} \tag{3.10}$$

Since only three of these are linearly independent we see that a generating matrix for the 1-cocycle space is

$$\begin{pmatrix} 0 & 1 & 0 & 1 & 1 & 0 \\ 0 & 0 & 1 & 0 & 1 & 1 \\ 1 & 1 & 1 & 0 & 0 & 0 \end{pmatrix} .$$

This code is orthogonal to the 1-cycle code.

4. Majority Decoding of Simplicial Codes

Let $t_0, t_1, \ldots, t_{n-1}$ denote the n binary digits of a word. A parity check for the digits is an equation of the form

$$c_0 t_0 + c_1 t_1 + \ldots + c_{n-1} t_{n-1} = 0 \tag{4.1}$$

where $c_0, c_1, \ldots, c_{n-1}$ are the integers 0 or 1 in GF(2). A digit t_j is said to be checked by a parity check equation if the coefficient of t_j in that equation is one. A set of parity check equations is said to be orthogonal on t_j if t_j is checked by every equation of the set and no other digit is checked by more than one equation of that set. Following Massey [2] we say that a block (n, k) code with minimum distance d is completely orthogonal if for each information digit t_j we can write $(d-1)$ parity check equations which are orthogonal on t_j $(j = 0, 1, \ldots, n-1)$.

Theorem. The p-cycle code defined by a simplex of dimension m is a completely orthogonal code with minimum weight $(p+2)$, and the p-cocycle code is a completely orthogonal code with minimum weight $(m-p+1)$ $(p = 1, 2, \ldots, (m-1))$.

To prove the assertion for p-cycle codes we shall exhibit a word of weight $p+2$ to guarantee that the minimum weight cannot exceed $p+2$ and then show that a set of $p+1$ orthogonal parity check equations can be constructed for each digit of the code words. Now Massey [2] has proved that if J parity checks can be formed which are orthogonal on a given digit of the code words of a linear code then the words of the code which differ in that digit are separated by a Hamming distance of at least $J+1$. It would follow from this that the minimum weight of the code is $(p+2)$.

The boundary of a $(p+1)$-simplex is a p-cycle. Since a $(p+1)$-simplex has $(p+2)$ vertices we see that the boundary of this simplex consists of $(p+2)$ p-faces. This shows that the minimum weight of the p-cycle code is no greater than $(p+2)$.

Consider the code word $t_0 t_1 \ldots t_{n-1}$. The vector y^p to which this word corresponds is

$$y^p = t_0 s_0^p + t_1 s_1^p + \ldots + t_{n-1} s_{n-1}^p .$$

To construct the parity check equations for t_q we form ρs_q^p, and use the coboundary of each $(p-1)$-simplex which occurs in ρs_q^p with a non-zero coefficient as a parity check equation. Let the vertices of s_q^p be denoted by $x_{j_0}^0, x_{j_1}^0, \ldots, x_{j_p}^0$ and let the remaining $(m-p)$ vertices of the original m-simplex be denoted by $x_{j_{p+1}}^0, x_{p+2}^0, \ldots, x_j^0$. Let the $(p-1)$-faces of s_q^p be denoted by $s_0^{p-1}, s_1^{p-1}, \ldots, s_p^{p-1}$ where

$$s_u^{p-1} = (x_{j_0}^0, x_{j_1}^0, \ldots, x_{j_{u-1}}^0, x_{j_{u+1}}^0, \ldots, x_{j_p}^0) .$$

The coboundary of this face is the sum of all the p-simplexes formed by adjoining one vertex of $(x_0^0, x_1^0, \ldots, x_m^0)$ which is not in s_u^{p-1}. Hence

$$\delta s_u^{p-1} = (x_{j_0}^0, x_{j_1}^0, \ldots, x_{j_p}^0)$$

$$+ \sum_{w=0}^{m-p-1} (x_{j_0}^0, x_{j_1}^0, \ldots, x_{j_{u-1}}^0, x_{j_{u+1}}^0, \ldots, x_{j_p}^0, x_{j_{p+1+w}}^0) \quad (u = 0, 1, 2, \ldots,$$

The $(p+1)$ parity-checks defined by these vectors all contain s_q^p with a non-zero coefficient and any two simplexes which correspond to distinct sets of values of u and w are distinct. Hence the (p+1 parity check equations defined by δs_u^{p-1} $(u = 0, 1, \ldots, p)$ are orthogonal on t_q. Since by the results of Section 2 coboundaries are parity checks for p-cycle code words, the proof of the part of the theorem concerning cycle codes is complete.

We consider next the p-cocycle codes. If s^{p-1} is a (p-1)-face of the original m-simplex then δs^{p-1} is a word of the cocycle code. But δs^{p-1} is the sum of all the p-faces of the original m-simplex which are formed by adjoining a vertex of the original m-simplex which is not in s^{p-1}. Hence δs^{p-1} consists of $[(m+1)-p]$ p-faces, and the corresponding word has the weight $(m+1)-p$. This shows that the minimum weight of the p-cocycle code cannot exceed $m+1-p$.

Consider the p-cocycle code word $t_0 t_1 \dots t_{n-1}$. The corresponding vector y^p is

$$y^p = t_0 s_0^p + t_1 s_1^p + \dots + t_{n-1} s_{n-1}^p .$$

To form the parity check equations for t_q we construct a set of $m-p$ p-cycles which contain s_q^p. Again, let the vertices of s_q^p be denoted by $x^0_{j_0}, x^0_{j_1}, \dots, x^0_{j_p}$, and let the remaining vertices of the original m-simplex be denoted by $x^0_{j_{p+1}}, x^0_{j_{p+2}}, \dots, x^0_{j_m}$.

Let

$$s_u^{p+1} = (x^0_{j_0}, x^0_{j_1}, \dots, x_{j_p}, x^0_{j_{p+1+u}}), \quad u = 0, 1, \dots, (m-p-1) .$$

Now

$$\rho s_u^{p+1} = (x^0_{j_0}, x^0_{j_1}, \dots, x^0_{j_p}) + \sum_{w=0}^{p} (x^0_{j_0}, \dots, x^0_{j_{w-1}}, x^0_{j_{w-1}}, \dots, x^0_{j_p}, x^0_{j_{p+1+u}}) .$$

By the results of Section 2 these p-cycles are parity checks for p-cocycle code words. Also, each parity check equation checks t_q, and these $(m-p)$ parity check equations are orthogonal on t_q since no two contain any p-simplex in common besides s_q^p. This completes the proof of the theorem.

To illustrate the construction of the orthogonal parity check equations consider the 1-cycle code in Section 3. To construct a check for the first digit of a code word, i.e., the coefficient of x_0^1, we note by (3.9),

$$\rho x_0^1 = \partial (x_0^0, x_3^0) = x_0^0 + x_3^0,$$

and by (3.10)

$$\delta x_0^0 = x_0^1 + x_3^1 + x_5^1 = 100101$$

$$\delta x_3^0 = x_0^1 + x_1^1 + x_2^1 = 111000 .$$

The last two words give the parity checks on the first digit of the 1-cycle code for a tetrahedron.

To get the parity check equations for the first digit of a word in the 1-cocycle code for a tetrahedron, note by (3.3).

$$x_0^1 = (x_0^0, x_3^0) \ .$$

Hence x_0^1 is in the boundaries of (x_0^0, x_3^0, x_1^0) and (x_0^0, x_3^0, x_2^0) . But by (3.3)

$$\rho(x_0^0, x_3^0, x_1^0) = (x_0^0, x_3^0) + (x_0^0, x_1^0) + (x_3^0, x_1^0) = x_0^1 + x_3^1 + x_1^1$$

$$\rho(x_0^0, x_3^0, x_2^0) = (x_0^0, x_3^0) + (x_0^0, x_2^0) + (x_3^0, x_2^0) = x_0^1 + x_5^1 + x_2^1 \ .$$

Hence the orthogonal parity checks on the first digit of the 1-cocycle codes are the two words

$$x_0^1 + x_1^1 + x_3^1 = 110100$$

$$x_0^1 + x_2^1 + x_5^1 = 101001 \ .$$

5. Code Parameters and the Euler - Poincaré Formula

The Euler-Poincaré formula will be used to compute the number of information digits for topological codes.

In the notation of Section 2 we define H^p as the quasi-orthogonal complement of ρV^{p+1} in Z^p, and $\bar{H}^p$ as the quasi-orthogonal complement of δV^{p-1} in $\bar{Z}^p$. Hence Z^p is the quasi-direct sum of H^p and ρV^{p+1}, and $\bar{Z}^p$ is the quasi-direct sum of $\bar{H}^p$ and δV^{p-1} ,

$$Z^p \sim H^p \oplus \rho V^{p+1} \ , \qquad (5.1)$$

$$\bar{Z}^p \sim \bar{H}^p \oplus \delta V^{p-1} \ . \qquad (5.2)$$

But by (2.13) and (2.14)

$$V^p \sim Z^p \oplus \delta V^{p-1} \sim H^p \oplus \rho V^{p+1} \oplus \delta V^{p-1} \qquad (5.3)$$

$$V^p \sim \bar{Z}^p \oplus \rho V^{p+1} \sim \bar{H}^p \oplus \delta V^{p-1} \oplus \rho V^{p+1} \ . \qquad (5.4)$$

A comparison of (5.3) and (5.4) shows that the spaces H^p and $\bar{H}^p$ have the same dimension. The common dimension B_p of H^p and $\bar{H}^p$ is called the pth Betti number of the complex K.

Let σ_p be the dimension of Z^p and let $\bar{\sigma}_p$ be the dimension of $\bar{Z}^p$. By (5.1)

$$\text{dimension } \rho V^{p+1} = \sigma_p - B_p \,. \tag{5.5}$$

Since the sum of the dimensions of the range ρV^{p+1} and the null space Z^{p+1} of the linear operator ρ restricted to V^{p+1} is the dimension of V^{p+1} we have,

$$\text{dimension } \rho V^{p+1} = r_{p+1} - \sigma_{p+1} \,. \tag{5.6}$$

Hence

$$r_{p+1} = \sigma_p + \sigma_{p+1} - B_p, \quad p = 0, 1, 2, \ldots \,. \tag{5.7}$$

Let μ_p be defined as

$$\mu_p = \sum_{j=0}^{p} (-1)^j r_j \,. \tag{5.8}$$

By (5.7)

$$\mu_p = r_0 + \sum_{j=1}^{p} (-1)^j r_j = r_0 + \sum_{j=1}^{p} (-1)^j (\sigma_{j-1} + \sigma_j - B_{j-1})$$

$$\mu_p = r_0 + \sum_{j=1}^{p} (-1)^{j+1} B_{j-1} - \sigma_0 + (-1)^p \sigma_p \,.$$

Thus

$$\sigma_p = (-1)^p (\mu_p + \sigma_0 - r_0 - \sum_{j=0}^{p-1} (-1)^j B_j) \,.$$

If we replace μ_p by its value in (5.8) we get

$$\sigma_p = r_p + (-1)^p [(\sigma_0 - r_0) + \sum_{j=0}^{p-1} (-1)^j (r_j - B_j)] \,.$$

Since the boundary of every 0-chain is 0 it follows that $\sigma_0 = r_0$ and therefore,

$$\sigma_p = r_p + (-1)^p \sum_{j=0}^{p-1} (-1)^j (r_j - B_j) . \tag{5.9}$$

This is a form of the Euler-Poincaré formula. The dimension, σ_p of the p-cycle space is the number of information digits of the p-cycle code and this is given by (5.9) in terms of the number of simplexes of dimension not exceeding p and the Betti numbers of the complex.

By (5.2),

$$\bar{\sigma}_p = B_p + \text{dimension } \delta V^{p-1}$$

and by (5.3),

$$r_p = B_p + \text{dimension } \delta V^{p-1} + \text{dimension } \rho V^{p+1} .$$

Hence by (5.5)

$$r_p = \bar{\sigma}_p + \sigma_p - B_p . \tag{5.10}$$

If this is written as

$$\bar{\sigma}_p = r_p + B_p - \sigma_p , \tag{5.11}$$

then in view of (5.9) we have the dimension of the p-cocycle space in terms of the number of simplexes of dimension not exceeding p and the Betti numbers of K .

6. Information Rates for Simplicial Codes

The complex K which is used in defining simplicial codes is an m-simplex S^m with vertices $x_0^0, x_1^0, \ldots, x_m^0$, together with all its faces. Thus every pair of vertices defines 1-simplex and the sum of every pair of vertices is a bounding cycle. Since a single vertex cannot be a boundary it follows that a 0-chain is a bounding cycle if and only if the number of non-zero coefficients in the 0-chain is even. Since the difference of any two 0-chains each of which has an odd number of non-zero coefficients is a 0-chain with an even number of non-zero coefficients, the 0-chains with an odd number of non-zero coefficients are a coset of the bounding 0-cycles. In particular, every element of this coset differs from x_0^0 by a bounding 0-chain. Since $V^0 = Z^0$ we see that

$$B_0 = \text{dimension } Z^0 - \text{dimension } \rho V^1 = 1 . \tag{6.1}$$

In addition, for $j > 1$, since every j-cycle is a bounding cycle (see Section 7(b)) we have

$$B_j = 0, \ j = 1, 2, \ldots \tag{6.2}$$

To determine r_j we note that the j-faces of S^m are all the subsets of $(j+1)$ of the $(m+1)$ vertices. Hence

$$r_j = \binom{m+1}{j+1} \quad j = 0, 1, 2, \ldots m . \tag{6.3}$$

An immediate consequence is the result that the word length n of both the p-cycle and p-cocycle codes is

$$n = r_p = \binom{m+1}{p+1} . \tag{6.4}$$

By (5.9), (6.1), and (6.2),

$$\begin{aligned} \sigma_p &= \binom{m+1}{p+1} + (-1)^p [-1 + \sum_{j=0}^{p-1} (-1)^j \binom{m+1}{j+1}] \\ \sigma_p &= (-1)^p \sum_{j=0}^{p+1} (-1)^{j-1} \binom{m+1}{j} \end{aligned} \tag{6.5}$$

We show by induction that

$$\sigma_p = \binom{m}{p+1} . \tag{6.6}$$

For $p = 0$, by (6.5) $\sigma_0 = -\binom{m+1}{0} + \binom{m+1}{1} = m$. Since $\binom{m}{1} = m$ the assertion is valid for $p = 0$. By (6.5)

$$\sigma_{p+1} = (-1)^{p+1} \sum_{j=0}^{p+2} (-1)^{j-1} \binom{m+1}{j} = -(-1)^p \sum_{j=0}^{p+1} (-1)^{j-1} \binom{m+1}{j} + \binom{m+1}{p+2} .$$

By the inductive assumption (6.6),

$$\sigma_{p+1} = -\binom{m}{p+1} + \binom{m+1}{p+2} = \binom{m}{p+2} .$$

This completes the proof of (6.6).

By (5.11) and (6.2),

$$\bar{\sigma}_p = r_p - \sigma_p = \binom{m+1}{p+1} - \binom{m}{p+1} = \binom{m}{p} \quad p = 1, 2, \ldots . \qquad (6.7)$$

This gives the number of information digits for the p-cocycle codes.

The code parameters for the simplicial codes are summarized in the following table.

	p-cycle codes	p-cocycle codes
n (word length)	$\binom{m+1}{p+1}$	$\binom{m+1}{p+1}$
k (number of information digits)	$\binom{m}{p+1}$	$\binom{m}{p}$
d (minimum weight)	$p+2$	$m-p+1$
$\frac{k}{n}$ (information rate)	$\frac{m-p}{m+1}$	$\frac{p+1}{m+1}$

7. Betti Numbers

a) Determination of Betti Numbers

By equations (2.3) and (2.4) we see that the rows of the incidence matrix span the space ρV^p. If we denote the pth incidence matrix [7] by I_p then

$$\text{dimension } \rho V^p = \text{rank } I_p . \qquad (7.1)$$

Similarly by Equation (2.5) we also have

$$\text{dimension } \delta V^{p-1} = \text{rank } I_p . \qquad (7.2)$$

Hence, by (5.6) we have

$$\sigma_{p+1} = r_{p+1} - \text{dimension } \rho V^{p+1} = r_{p+1} - \text{rank } I_{p+1} . \qquad (7.3)$$

Similarly,

$$\sigma_p = r_p - \text{rank } I_p \ . \tag{7.4}$$

If we substitute for σ_p and σ_{p+1} in (5.7) and solve for B_p we get

$$B_p = r_p - \text{rank } I_p - \text{rank } I_{p+1} \ . \tag{7.5}$$

This gives the $p^{\underline{th}}$ Betti number of a complex K in terms of the number of p-simplexes in K and the ranks of the $p^{\underline{th}}$ — and $(p+1)^{\underline{st}}$ — incidence matrices for K .

b) <u>The Betti Numbers of a Simplex</u>

In this section we show that the Betti numbers $B_p (0 < p < m)$ of an m-simplex are all zero by showing that every p-cycle is a boundary. Let K be the complex consisting of the m-simplex $\{x_0^0, x_1^0, \ldots, x_m^0\}$ and all of its faces.

The following remarks will be needed in the proof. If x_0^q is the q-simplex $\{x_{i_0}^0, x_{i_1}^0, \ldots, x_{i_q}^0\}$ then $\{x_{i_{q+1}}^0, x_0^q\}$ will denote the $(q+1)$-simplex $\{x_{i_{q+1}}^0, x_{i_0}^0, x_{i_1}^0, \ldots, x_{i_q}^0\}$. We shall also make the definition,

$$\{x_i^0, ax_j^q + bx_k^q\} = a\{x_i^0, x_j^q\} + b\{x_i^0, x_k^q\} \ . \tag{7.6}$$

Thus

$$\rho\{x_i^0, x_j^q\} = \{x_j^q\} + \{x_i^0, \rho x_j^q\} = x_j^q + \{x_i^0, \rho x_j^q\} \ . \tag{7.7}$$

Let y_1^p be a p-cycle,

$$y_1^p = \sum_{j=0}^{r_p - 1} a_j x_j^p \ . \tag{7.8}$$

We write this cycle as the sum of two p-chains, one in which the p-simplexes do not have x_m^0 as a vertex and one in which the p-simplexes have x_m^0 as a vertex. The number of p-simplexes which do not have x_m^0 as a vertex is

$$c = \binom{m}{p+1} \ . \tag{7.9}$$

Thus, if we let $x_0^p, x_1^p, \ldots, x_{c-1}^p$ be the p-simplexes which do not have x_m^0 as a vertex, and if we set $x_{i+c}^p = \{x_m^0, x_i^{p-1}\}$, $i = 0, 1, \ldots, (d-1)$, where $d = \binom{m}{p}$ and x_i^{p-1} are the (p-1)-faces of $\{x_0^0, x_1^0, \ldots, x_{m-1}^0\}$, we have

$$y_1^p = \sum_{j=0}^{c-1} a_j x_j^p + \sum_{i=0}^{d-1} a_{i+c}\{x_m^0, x_i^{p-1}\} . \qquad (7.10)$$

Let

$$y_2^p = \rho(\sum_{j=0}^{c-1} a_j\{x_m^0, x_j^p\}) . \qquad (7.11)$$

Then y_2^p is a p-cycle and

$$y_2^p = \sum_{j=0}^{c-1} a_j x_j^p + \sum_{j=0}^{c-1} a_j\{x_m^0, \rho x_m^p\} . \qquad (7.12)$$

It follows that $y_1^p + y_2^p$ is a p-cycle and

$$y_1^p + y_2^p = \sum_{i=0}^{d-1} a_{i+c}\{x_m^0, x_i^{p-1}\} + \sum_{j=0}^{c-1} a_j\{x_m^0, \rho x_j^p\} . \qquad (7.13)$$

By expanding the second sum we see there are coefficients $b_0, b_1, \ldots, b_{d-1}$ such that

$$y_1^p + y_2^p = \sum_{j=0}^{d-1} b_j \{x_m^0, x_j^{p-1}\} \qquad (7.14)$$

where the simplexes x_j^{p-1} $(j = 0, 1, \ldots, d-1)$ do not have x_m^0 as a vertex. Hence,

$$0 = \rho(y_1^p + y_2^p) = \sum_{j=0}^{d-1} b_j x_j^{p-1} + \sum_{j=0}^{d-1} b_j \{x_m^0, \rho x_j^{p-1}\} . \qquad (7.15)$$

Since each (p-1)-simplex of the last sum is not a simplex which occurs in the first sum we must have

$$\sum_{j=0}^{d-1} b_j x_j^{p-1} = 0 . \qquad (7.16)$$

But the simplexes $x_0^{p-1}, x_1^{p-1}, \ldots, x_{d-1}^{p-1}$ are linearly independent. Therefore $b_0 = b_1 = \ldots = b_{d-1} = 0$. It follows from (7.14) that $y_1^p + y_2^p = 0$ or

$$y_1^p = y_2^p \tag{7.17}$$

By (7.11) we see that y_1^p is a bounding p-cycle. This shows that every p-cycle $(0 < p < m)$ on an m-simplex is a boundary and therefore $B_p = 0 (0 < p < m)$.

REFERENCES

1. L. S. Pontryagin, Foundations of Combinatorial Topology, Graylock Press (1952).

2. James L. Massey, Threshold Decoding, M.I.T. Press (1963)

3. R. L. Townsend and E. J. Weldon, "Self-Orthogonal Quasi-Cyclic Codes", IEEE Trans. on Information Theory, Vol. IT-13, pp. 183-195, April 1967.

4. L. D. Rudolph, "A Class of Majority Logic Decodable Codes", IEEE Trans. on Information Theory, Vol. IT-13, pp. 305-307, April 1962.

5. D. A. Huffman, "A Graph-Theoretic Formulation of Binary Group Codes", Summaries presented at ICMCI, Part 3, Sept. 1964, pp. 29-30.

6. S. L. Hakimi and H. Frank, "Cut-Set Matrices and Linear Codes", IEEE Trans. on Information Theory, Vol. IT-11, pp. 457-458, July 1965.

7. Beno Eckmann, "Harmonische Funktionen und Randwertaufgaben in einem Komplex", Commentarii Mathematici Helvetici, Vol. 17, 1944-5, pp. 240-255.

Sponsored in part by the National Science Foundation through Grant GN 534 from the Office of Science Information Service to the Computer and Information Science Research Center, The Ohio State University and the Ohio State University Research Fund.

S. LIN

On a Class of Cyclic Codes

ABSTRACT

Two subclasses of polynomial codes have been studied. One subclass of polynomial codes has been proved to contain the Euclidean geometry codes as a proper subclass. Another subclass of polynomial codes has been shown to be closely related to the projective geometry codes. A BCH lower bound on the minimum distance of Euclidean geometry codes has been derived.

1. Introduction

The polynomial approach of Kasami, et al, to a class of cyclic codes (polynomial codes) [1, 2] gives a unified formulation for several important classes of codes, and puts the latter codes into a large framework. Two subclasses of polynomial codes are closely related to the geometry codes which have been studied extensively for the past few years [3, 4, 5, 6, 7, 8, 9, 10, 11, 12, 13, 14, 15, 16, 17]. It is the purpose of this paper to establish the relationship between polynomial codes and geometry codes. Firstly, we shall summarize the important results of polynomial codes. Secondly, we shall prove that one subclass of polynomial codes contains the Euclidean geometry (E. G.) codes [7] and the generalized Reed-Muller codes [6, 10, 12] as subclasses. Thirdly, a relation between one subclass of polynomial codes and finite projective geometry codes [8, 9] will be established. A BCH lower bound on the minimum distance of Euclidean geometry codes is derived. From this BCH bound, we are able to show that E. G. codes are in general more powerful than Weldon's decoding scheme has been able to demonstrate.

This work is based primarily on the results in the first and second references. Where possible the notation and conventions employed therein will be followed here.

2. Summarized Results on Polynomial Codes [1, 2]

Let q be a power of prime, say $q = p^c$, and α be a primitive root of $GF(q^{ms})$. Any non-zero element α^j in $GF(q^{ms})$ can be expressed as

$$\alpha^j = \sum_{i=1}^{m} a_{ij}\,\alpha^{i-1} \quad \text{for} \quad 0 \leq j < q^{ms} - 1 \,, \tag{1}$$

where $a_{ij} \in GF(q^s)$. The correspondence between α^j and the m-tuple $(a_{1j}, a_{2j}, \ldots, a_{mj})$ is one-to-one.

Let b be a factor of $q^s - 1$, and

$$z = (q^s - 1)/b \,, \qquad n = (q^{ms} - 1)/b \,. \tag{2}$$

Let $X_1, X_2, \ldots, X_m$ be m variables over $GF(q^s)$ and $\bar{X} = (X_1, X_2, \ldots, X_m)$. Define

$$P(m, s, \mu, b) \tag{3}$$

as the set of polynomials $f(\bar{X}) = f(X_1, X_2, \ldots, X_m)$ in $X_1, X_2, \ldots, X_m$ with coefficients in $GF(q^s)$ such that the sum of the exponents of each term of $f(\bar{X})$ is a multiple of b and the degree of $f(\bar{X})$ is μb or less. Therefore, each polynomial in $P(m, s, \mu, b)$ is of the following form

$$f(\bar{X}) = \sum C X_1^{\nu_1} X_2^{\nu_2} \cdots X_m^{\nu_m} \tag{4}$$

where $C \in GF(q^s)$, $0 \leq \nu_i < q^s$ and $\sum_{i=1}^{m} \nu_i = jb$ with $0 \leq j \leq \mu$. The parameter μ is at most mz.

Define a vector

$$\bar{v}(f) = (v_0, v_1, v_2, \ldots, v_{n-1}) \tag{5}$$

whose components are in $GF(q^s)$ as follows

$$v_j = f(a_{1j}, a_{2j}, \ldots, a_{mj}) \tag{6}$$

for $0 \leq j < n$, where $f(\bar{X})$ is in $P(m, s, \mu, b)$, $(a_{1j}, a_{2j}, \ldots, a_{mj}) \Longleftrightarrow \alpha^j$, and $n = (q^{ms} - 1)/b$.

Now, define

$$Q(m, s, \mu, b, q) \tag{7}$$

to be the set of all polynomials $f(\bar{X})$ in $P(m, s, \mu, b)$ such that

$$f(a_{1j}, a_{2j}, \ldots, a_{mj}) \in GF(q) \tag{8}$$

for $0 \leq j < n$.

Definition [Kasami, Lin and Peterson]: A q-ary (n, m, s, μ, q)-polynomial code is defined as the set of vectors $\bar{v}(f)$

$$\{\bar{v}(f) \mid f(\bar{X}) \in Q(m, s, \mu, b, q)\} \ . \tag{9}$$

Let h be a non-negative integer less than q^{ms} . Express h in radix-q^s form as follows

$$h = \delta_0 + \delta_1 q^s + \delta_2 q^{2s} + \ldots + \delta_{m-1} q^{(m-1)s} \tag{10}$$

where $0 \leq \delta_i < q^s$ for $0 \leq i < m$. The q^s-weight of h is defined as

$$W_{q^s}(h) = \sum_{i=0}^{m-1} \delta_i \ . \tag{11}$$

Let

$$h' \equiv hq^{\ell} \ (\text{mod } q^{ms} - 1) \ . \tag{12}$$

Then the q^s-weight of h' is defined as

$$W_{q^s}(h') = W_{q^s}(hq^{\ell}) \ . \tag{13}$$

Kasami, et al, proved the following two theorems:

Theorem 1: A q-ary (n, m, s, μ, q)-polynomial code is a cyclic code which has the following parameters:

a) Code length $n = (q^{ms}-1)/b$

b) Number of information digits

$$k = \left\{ \begin{array}{l} \text{the number of non-negative integers } h \text{ less} \\ \text{than } q^{ms}-1 \text{ which are divisible by } b \text{ and such that} \\ \max\limits_{0 \leq \ell < s} W_{q^s}(hq^{\ell}) = jb \text{ with } 0 \leq j \leq \mu \ . \end{array} \right\}$$

c) Minimum distance d_{min}

$$d_{min} \geq [\,(R+1)\,q^{Qs}-1]/b$$

where Q and R are quotient and remainder resulting from dividing $m(q^s-1)-\mu b$ by q^s-1. The generator polynomial $g(X)$ has α^j as a root if and only if there exists an h such that h is divisible by b, and

$$\min_{0 \leq \ell < s} W_{q^s}(hq^{\ell}) = jb$$

with $0 < j < mz - \mu$.

Theorem 2. The q-ary dual code of a (n, m, s, μ, q)-polynomial code has the following parameters:

a) $n = (q^{ms}-1)/b$.

b) Number of information digits

$$k^0 = \begin{cases} \text{the number of non-negative integers } h \text{ less} \\ \text{than } q^{ms}-1 \text{ which are divisible by } b \text{ and such that} \\ \min_{0 \leq \ell < s} (hq^{\ell}) = jb \text{ with } 0 < j < mz - \mu . \end{cases}$$

The generator polynomial $g^0(X)$ has α^h as a root if and only if h is divisible by b and

$$\max_{0 \leq \ell < s} W_{q^s}(hq^{\ell}) = jb$$

with $0 \leq j \leq \mu$.

A general lower bound on the minimum distance of the dual code of a (n, m, s, μ, q)-polynomial code has not been obtained. But, for some special cases, we are able to derive a BCH lower bound.

3. A Subclass of Polynomial Codes Which is Related to Euclidean Geometry Codes

The class of cyclic codes based on the Euclidean geometry was first introduced by Rudolph [3], and then has been studied extensively by Weldon and others [7, 11, 14, 16, 17]. Codes of this class are called Euclidean geometry codes. In general, a code in this class is less efficient than the corresponding BCH code of the same designed minimum distance. But it can be decoded with a relatively modest amount of equipment. In this section, we shall prove that dual codes of the class of polynomial codes with $b = 1$ contain the E. G. codes as a subclass.

For $b = 1$, a (n, m, s, μ, q)-polynomial code has the following parameters:

$$n = q^{ms} - 1 ,$$

$$k = \left\{ \begin{array}{l} \text{the number of non-negative integers } h \text{ less} \\ \text{than } q^{ms}-1 \text{ such that } \max\limits_{0 \le \ell < s} W_{q^s}(hq^{\ell}) = \mu \end{array} \right\} ,$$

$$d_{min} \geq (q^s - N) q^{(m-D-1)s} - 1$$

where D and N are quotient and remainder resulting from dividing μ by q^s-1, i.e.

$$\mu = D(q^s-1) + N \tag{14}$$

with $0 \leq N < q^s - 1$. The generator polynomial $g(X)$ has α^h as a root if and only if

$$0 < \min_{0 \leq \ell < s} W_{q^s}(hq^{\ell}) < m(q^s-1) - \mu . \tag{15}$$

For $s = 1$, this subclass of polynomial codes reduces to generalized Reed-Muller codes [10]. Thus, we may consider this subclass as a new generalization of Reed-Muller codes.

Let L be a t-dimensional subspace of all m-tuples over $GF(q^s)$ and $\{A_1, A_2, \ldots, A_t\}$ be a basis of L . Let Γ be the null space of L . Then the dimension of Γ is $m - t$. The following polynomial

$$f_L(\bar{X}) = \prod_{i=1}^{t} \{1 - (A_t \cdot \bar{X}^T) q^s - 1\} \tag{16}$$

has degree $t(q^s - 1)$, where $A_t \cdot \bar{X}^T$ is the inner product of A_t and $\bar{X}$, i.e.

$$A_t \cdot \bar{X}^T = (a_{1t}, a_{2t}, \ldots, a_{mt})(X_1, X_2, \ldots, X_m)^T$$

$$= \sum_{i=1}^{m} a_{it} X_i \quad (\text{over } GF(q^s)) .$$

Express μ as in Equation (14). For $0 \leq t \leq D$, $f_L(\bar{X})$ is a polynomial in $Q(m, s, \mu, 1, q)$ and vector $\bar{v}(f)$ defined in accordance with Equation (5) and Equation (6) is a code vector in the $(q^{ms}-1, m, s, \mu, q)$-polynomial code. Since

$$f_L(\bar{X}_j) = 1 \text{ for } X_j \in \Gamma$$

and

$$f_L(\bar{X}_j) = 0 \text{ for } \bar{X}_j \notin \Gamma$$

the components of $\bar{v}(f_L)$ are "1" at the locations corresponding to the non-zero m-tuples of Γ and "0" at other locations. The Hamming weight of $\bar{v}(f_L)$ is $q^{(m-t)s}-1$. Let Γ' be a coset with respect to the subspace Γ. Then, the polynomial

$$f'_L(\bar{X}) = \prod_{i=1}^{t} \{1 - [A_i \cdot (\bar{X} - B)^T]^{q^s-1}\} \tag{17}$$

is also in $Q(m, s, \mu, 1, q)$ for $0 \leq t \leq D$, where B is any vector in Γ'. Therefore, the components of $\bar{v}(f'_L)$ are "1" at the locations corresponding to the elements of Γ' and are "0" at other locations. The weight of $\bar{v}(f'_L)$ is $q^{(m-t)s}$.

All the m-tuples over $GF(q^s)$ form a Euclidean geometry over $GF(q^s)$, i.e. $E.G.(m, q^s)$. Every m-tuple is a point in $E.G.(m, q^s)$. The all zero m-tuple is regarded as the point at infinity. Therefore, every bit position of a code vector in a $(q^{ms}-1, m, s, \mu, q)$-polynomial code can be uniquely associated with a point in $E.G.(m, q^s)$ except the point at infinity. Γ is a $(m-t)$-flat through the point at infinity, and Γ' is a $(m-t)$-flat through the point B. For convenience, we call the code vector $\bar{v}(f_L)$ or $\bar{v}(f'_L)$ a $(m-t)$-flat in $E.G.(m, q^s)$. By Equation (16) and Equation (17), we obtain:

<u>Theorem 3</u>. For $0 \leq t \leq D$, every $(m-t)$-flat of $E.G.(m, q^s)$ is a code vector in a $(q^{ms}-1, m, s, \mu, q)$-polynomial code, where $\mu = D(q^s-1) + N$ with $0 \leq N < q^s-1$.

Thus, a $(q^{ms}-1, m, s, \mu, q)$-polynomial code contains all the $(m-D)$-flats, $(m-D+1)$-flats, ..., and m-flats of $E.G(m, q^s)$. The smallest flats which are contained in the code are $(m-D)$-flats. The number of $(m-D)$-flats which intersect on a given $(m-D-1)$-flat is

$$\begin{aligned} J &= 1 + q^s + q^{2s} + \dots + q^{Ds} \\ &= \frac{q^{(D+1)s} - 1}{q^s - 1} . \end{aligned} \tag{18}$$

It follows from Weldon's argument [7], that the dual code of a $(q^{ms}-1, m, s, \mu, q)$-polynomial code is $(m-D)$-step orthogonalizable [18] and has minimum distance at least

$$J + 1 = \frac{q^{(D+1)s}-1}{q^s - 1} + 1 \tag{19}$$

where $\mu = D(q^s - 1) + N$ with $0 \leq N < q^s - 1$. Equation (19) gives a lower bound for the minimum distance of the dual code of a $(q^{ms}-1, m, s, \mu, q)$-polynomial code. For large D and S, this bound is very loose. A BCH lower bound can be derived by counting the consecutive roots in the generator polynomial of the dual code.

From Theorem 2, the generator polynomial $g^0(X)$ of the dual of a $(q^{ms}-1, m, s, \mu, q)$-polynomial code has α^h as a root if and only if

$$\max_{0 \leq \ell < s} W_{q^s}(hq^{\ell}) \leq \mu .$$

Let $\mu = D(q^s - 1) + N$ with $0 \leq N < q^s - 1$. Consider the following integer

$$\begin{aligned} h_0 &= (q^s - 1) + (q^s - 1)q^s + \ldots + (q^s - 1)q^{(D-2)s} \\ &\quad + (q-1)q^{(D-1)s} + (\lambda + 1)q^{Ds} \end{aligned} \tag{20}$$

where λ is the quotient resulting from dividing N by q^{s-1}, i.e.

$$N = \lambda q^{s-1} + \sigma \tag{21}$$

with $0 \leq \sigma < q^{s-1}$. Since $N < q^s - 1$, therefore, $0 \leq \lambda \leq q - 1$. For $0 \leq \ell < s$, the radix-q^s expansion of $h_0 q^{\ell}$ is

$$\begin{aligned} h_0 q^{\ell} &= (q^s - q^{\ell}) + (q^s - 1)q^s + \ldots + (q^s - 1)q^{(D-2)s} \\ &\quad + (q^{\ell+1} - 1)q^{(D-1)s} + (\lambda + 1)q^{\ell} q^{Ds} . \end{aligned} \tag{22}$$

The q^s-weight of $h_0 q^{\ell}$ is

$$W_{q^s}(h_0 q^{\ell}) = (D-1)(q^s - 1) + (\lambda + q)q^{\ell} . \tag{23}$$

It is obvious that

$$\max_{0 \le \ell < s} W_{q^s}(h_0 q^\ell) = W_{q^s}(h_0 q^{s-1}) = D(q^s - 1) + \lambda q^{s-1} + 1 . \tag{24}$$

It is easy to show that any integer h less than h_0 satisfies

$$\max_{0 \le \ell < s} W_{q^s}(hq^\ell) \le D(q^s - 1) + \lambda q^{s-1} + 1 . \tag{25}$$

If N is divisible by q^{s-1} $(\sigma = 0)$, then h_0 is the smallest integer such that

$$\max_{0 \le \ell < s} W_{q^s}(h_0 q^\ell) = D(q^s - 1) + \lambda q^{s-1} + 1 = \mu + 1 . \tag{26}$$

If N is not divisible by q^{s-1} $(\sigma \ne 0)$, then

$$\max_{0 \le \ell < s} W_{q^s}(h_0 q^\ell) \le \mu . \tag{27}$$

By Equation (25), Equation (26) and Equation (27), the dual code of a $(q^{ms}-1, m, s, \mu, q)$-polynomial code has at least the following consecutive roots

$$\alpha^0, \alpha^1, \alpha^2, \ldots, \alpha^{h_0 - 1} \tag{28}$$

where α is a primitive root of $GF(q^{ms})$. By Bose's argument [19], the dual code has minimum distance at least

$$h_0 + 1 = (\lambda+1) q^{Ds} + q \cdot q^{(D-1)s} .$$

<u>Theorem 4</u>. The dual code of a $(q^{ms}-1, m, s, \mu, q)$-polynomial code has minimum distance at least

$$(\lambda+1) q^{Ds} + q \cdot q^{(D-1)s} \tag{29}$$

where D and λ satisfy the following equations

$$\mu = D(q^s - 1) + N \qquad 0 \le N < q^s - 1 ,$$

$$N = \lambda q^{s-1} + \sigma \qquad 0 \le \sigma < q^{s-1} .$$

For $\mu = D(q^s - 1) + N$ with $0 \le N < q^s - 1$, Theorem 3 tells us that the smallest flats of $E.G.(m, q^s)$ are $(m-D)$-flats. By Weldon's argument, the dual code of a $(q^{ms}-1, m, s, \mu, q)$-polynomial code is (m-D)-step orthogonalizable. The polynomial code of the smallest dimension which contains (m-D)-flats is the code with $\mu = D(q^s-1)$, i.e. a $(q^{ms}-1, m, s, D(q^s-1), q)$-polynomial code. Therefore, the dual code of a $(q^{ms}-1, m, s, D(q^s-1), q)$-polynomial code is the code of the largest dimension which contains (m-D)-flats of E.G. (m, q^s) in its null space. If Weldon's decoding scheme is used, the dual code of a $(q^{ms}-1, m, s, \mu, q)$-P-code with $\mu = D(q^s-1) + N$ and the dual code of a $(q^{ms}-1, m, s, D(q^s-1), q)$-P-code will correct the same number of errors $\left\lfloor \frac{J}{2} \right\rfloor$*, where

$$J = \frac{q^{(D+1)s}-1}{q^s-1} \quad . \tag{30}$$

In the following, we shall prove that, for $q = 2$, the dual code of a $(2^{ms}-1, m, s, D(2^s-1), 2)$-polynomial code is an E.G. code defined by Weldon [7]. Let h be a positive integer less than $2^{ms}-1$. In radix-2^s expansion

$$h = \sum_{i=0}^{m-1} \delta_i 2^{is} \tag{31}$$

where $0 \leq \delta_i < 2^s$ for $0 \leq i \leq m-1$.
In radix-2 expansion,

$$h = \sum_{i=0}^{m-1} \sum_{t=0}^{s-1} \delta_{it} 2^{t+is} \tag{32}$$

where $\delta_{ij} = 0$ or 1 for $0 \leq i \leq m-1$ and $0 \leq t \leq s-1$. The binary ms-tuple associated with h may contain some disjoint binary representations of multiples of 2^s-1. Now define the s-weight of h, denoted by $W_s(h)$, as the maximum number of such disjoint multiples [7].

Definition (Weldon) [7]. The $(\nu, s)^{th}$ order cyclic E.G. code is a code whose parity check polynomial $h(X)$ contains among its roots all α^h such that $W_s(h) \leq \nu$.

Theorem 5. For a positive integer h less than $2^{ms}-1$, $W_s(h) \leq \nu$ if and only if $\min_{0 \leq \ell < s} W_{2^s}(h2^\ell) < (\nu+1)(2^s-1)$.

Proof: Step 1. We show that $\min_{0 \leq \ell < s} W_{2^s}(h2^\ell) < (\nu+1)(2^s-1) \Longrightarrow W_s(h) \leq \nu$.

---

* The symbol $\lfloor y \rfloor$ denotes the greatest integer contained in y.

Let

$$h = h_0 + h_1 + h_2 + \ldots + h_r \tag{33}$$

where

a) $h_0 \geq 0$ and $h_i > 0$ for $1 \leq i \leq r$,

b) h_i is a multiple of 2^s-1, for $1 \leq i \leq r$,

and c) the radix-2 expansions of $h_1, h_2, \ldots, h_r$ are mutually disjoint.

Then,

$$W_{2^s}(h_j 2^\ell) = W_{2^s}(h_0 2^\ell) + W_{2^s}(h_1 2^\ell) + \ldots + W_{2^s}(h_r 2^\ell) . \tag{34}$$

Since $h_i 2^\ell$ is divisible by 2^s-1, $W_{2^s}(h_i 2^\ell)$ is a multiple of 2^s-1. Let $W_{2^s}(h_i 2^\ell) = k_i(2^s-1)$, for $1 \leq i \leq r$, where $k_i > 0$. Then

$$\begin{aligned} W_{2^s}(h2^\ell) &= W_{2^s}(h_0 2^\ell) + (2^s-1)\sum_{i=1}^{r} k_i \\ &\geq (2^s-1)\sum_{i=1}^{r} k_i \qquad (35) \\ &\geq r(2^s-1) . \end{aligned}$$

Therefore, we have

$$\nu + 1 > r . \tag{36}$$

Since the inequality is true for any possible r, thus

$$\min_{0 \leq \ell < s} W_{2^s}(hq^\ell) < (\nu+1)(2^s-1) \Longrightarrow W_s(h) \leq \nu .$$

Step 2 . We show that $W_s(h) \leq \nu \Longrightarrow \min_{0 \leq \ell < s} W_{2^s}(h2^\ell) < (\nu+1)(2^s-1)$.

Let $W_s(h) = j$ where $j \leq \nu$. Then h can be expressed as a sum follows

$$h = h_0 + h_1 + \ldots + h_j \tag{37}$$

where a) $h_0 \geq 0$ and $h_i > 0$ for $1 \leq i \leq j$,

b) h_i is a multiple of 2^s-1, for $1 \leq i \leq j$,

c) The radix-2 expansions of $h_1, \ldots, h_j$ are mutually disjoint.

By the definition of s-weight, the 2^s-weight of h_i must be equal to 2^s-1 exactly,

$$W_{2^s}(h_i) = (2^s-1)$$

for $1 \leq i \leq j$. Also $W_{2^s}(h_0) < 2^s-1$. Otherwise, h_0 can be partitioned into at least two disjoint parts such that one part is divisible by 2^s-1 . This is a contradiction to the assumption that j is the maximum number of disjoint multiples of (2^s-1) . Therefore, we obtain

$$W_{2^s}(h) < (j+1)(2^s-1) . \tag{38}$$

Equation (38) implies that $W_{2^s}(h) < (\nu+1)(2^s-1)$ if $W_s(h) \leq \nu$. The theorem is then proved. Q.E.D.

Theorem 5 gives a relation between E.G. codes and polynomial codes. By Theorem 1, the definition of Weldon's E.G. codes and Theorem 5, we have

Theorem 6. The dual code of a binary $(2^{ms}-1, m, s, D(2^s-1), 2)$-polynomial code with an overall parity check ($\alpha^0=1$ as a root in its generator polynomial) is a $[m-D-1, s]^{th}$ order E.G. code.

For $\mu = D(q^s-1)$, we may consider the dual of a $(q^{ms}-1, m, s, D(q^s-1), q)$-P-code with an overall parity check as a generalized $(m-D-1, s)^{th}$ order Euclidean geometry code.

Weldon's decoding scheme only demonstrates that a $(m-D-1, s)^{th}$ order E.G. code can correct $\left\lfloor \frac{J}{2} \right\rfloor$ errors, where

$$J = \frac{q^{(D+1)s}-1}{q^s-1} . \tag{39}$$

Theorem 4 tells us that a q-ary $(m-D-1, s)^{th}$ order E.G. code has error correcting capability at least $\left\lfloor \frac{h_0-1}{2} \right\rfloor$ where

$$h_0 = q^{Ds} + q \cdot q^{(D-1)s} - 1 \tag{40}$$

The difference is

$$(q-2)q^{(D-1)s} + (q^s-2)q^{(D-2)} + \ldots + (q^s-2) . \tag{41}$$

Therefore, a $(\nu, s)^{th}$ order E.G. code is, in general, more powerful than Weldon's decoding scheme has been able to demonstrate.

Consider an integer less than $q^{ms}-1$

$$h = \delta_0 + \delta_1 q^s + \ldots + \delta_{m-1} q^{(m-1)s} .$$

where $0 \leq \delta_i < q^s$ for $0 \leq i \leq m-1$. We define $h' = \delta'_0 + \delta'_1 q^s + \ldots + \delta'_{m-1} q^{(m-1)s}$ as a descendant of h if and only if

$$0 \leq \delta'_i \leq \delta_i \quad \text{for} \quad 0 \leq i \leq m-1 .$$

The necessary and sufficient condition for an argumented primitive cyclic code of length q^{ms} to be invariant under the affine group of permutations is that, for every α^h which is a root of the generator polynomial, $\alpha^{h'}$ is also a root, where α is a primitive element of $GF(q^{ms})$ [6]. By Theorem 1, the argumented $(q^{ms}-1, m, s, \mu, q)$-P-code is doubly-transitive invariant under the affine group of permutations.

4. A Subclass of Polynomial Codes Which is Related to the Projective Geometry Codes

Projective geometry codes were first introduced by L. D. Rudolph [3] in 1964. Since then, this class of codes have been studied estensively by many authors, especially by Weldon [4, 7, 17] and Goethals et al, [9]. In this section, we shall prove the relationship between projective geometry codes and a certain subclass of polynomial codes.

For $b = q^s-1$, a q-ary $(\frac{q^{ms}-1}{q^s-1}, m, s, \mu, q)$-polynomial code has the following parameters

$$n = \frac{q^{ms}-1}{q^s-1} ,$$

$$k = \left\{ \begin{array}{l} \text{The number of non-negative integers } h \text{ less} \\ \text{than } q^{ms}-1 \text{ which is divisible by } q^s-1 \text{ and} \\ \max_{0 \leq \ell < s} W_{q^s}(hq^\ell) = j(q^s-1) \text{ with } 0 \leq j \leq \mu . \end{array} \right\} ,$$

$$d = [q^{(m-\mu)s}-1]/q^s-1 .$$

The generator polynomial has α^h as a root if and only if

$$\min_{0 \le \ell < s} W_{q^s}(hq^\ell) = j(q^s - 1)$$

with $0 < j < m - \mu$.

In the following, we shall prove the relationship between a $(\frac{q^{ms}-1}{q^s-1}, m, s, \mu, q)$-polynomial code and projective geometry code.

Let h be a non-negative integer which is less than $q^{ms}-1$. Express h in both radix-q^s form and radix-q form as follows:

(1) Radix-q^s form

$$h = \sum_{i=0}^{m-1} \delta_i q^{is} \tag{42}$$

where $0 \le \delta_i < q^s$ for $0 \le i \le m-1$,

(2) Radix-q form

$$h = \sum_{i=0}^{m-1} \sum_{t=0}^{s-1} \delta_{it} q^{t+is} \tag{43}$$

where $0 \le \delta_{it} < q$ for $0 \le i \le m-1$ and $0 \le t \le s-1$.

The q^s-weight of h is $W_{q^s}(h) = \sum_{i=0}^{m-1} \delta_i$ and the q-weight of h is $W_q(h) = \sum_{i=0}^{m-1} \sum_{t=0}^{s-1} \delta_{it}$.

Theorem 7:

$$(q^s - 1) W_q(h) = (q-1) \sum_{\ell=0}^{s-1} W_{q^s}(hq^\ell) \ . \tag{44}$$

Proof. Let

$$\sigma_t = \sum_{i=0}^{m-1} \delta_{it} \ . \tag{45}$$

Then,

$$W_{q^s}(h) = \sum_{t=0}^{s-1} \sigma_t q^t \ . \tag{46}$$

For any $0 \leq \ell < s$, we have

$$W_{q^s}(hq^\ell) = q^\ell \sum_{t=0}^{s-\ell-1} \sigma_t q^t + q^{-(s-\ell)} \sum_{t=s-\ell}^{s-1} \sigma_t q^t . \tag{47}$$

Then

$$\sum_{\ell=0}^{s-1} W_{q^s}(hq^\ell) = \sum_{\ell=0}^{s-1} q^\ell \sum_{t=0}^{s-\ell-1} \sigma_t q^t + \sum_{\ell=0}^{s-1} q^{-(s-\ell)} \sum_{t=s-\ell}^{s-1} \sigma_t q^t$$

$$= \frac{q^s-1}{q-1} \sum_{t=0}^{s-1} \sigma_t$$

$$= \frac{q^s-1}{q-1} W_q(h) . \tag{48}$$

Rearranging Equation (48), we obtain

$$(q^s-1) W_q(h) = (q-1) \sum_{\ell=0}^{s-1} W_{q^s}(hq^\ell) . \tag{49}$$

Q.E.D.

Corollary 8

$$(q^s-1) W_q(h) \geq s(q-1) \min_{0 \leq \ell < s} W_{q^s}(hq^\ell) \tag{50}$$

and

$$(q^s-1) W_q(h) \leq s(q-1) \max_{0 \leq \ell < s} W_{q^s}(hq^\ell) . \tag{51}$$

Definition (Weldon) [8]. The binary μs^{th} order projective geometry code (or non-primitive Reed-Muller code as it was called) is a cyclic code of length $\frac{2^{ms}-1}{2^s-1}$ whose generator polynomial contains among its roots all α^h such that h is divisible by 2^s-1 and

$$W_2(h) \leq \mu s . \tag{52}$$

From Equation (50) and Equation (51), it is obvious that

$$W_2(h) \leq \mu s \Longrightarrow \mu(2^s-1) \geq \min_{0 \leq \ell < s} W_{2^s}(h2^\ell) \tag{53}$$

where he is divisible by 2^s-1 .

Equation (53) implies the following theorem

Theorem 9: A binary $(\frac{2^{ms}-1}{2^s-1}, m, s, m-\mu-1, 2)$-polynomial code with α^0 in the generator polynomial is a subcode of a μs order binary P.G. code.

By Theorem 2, the generator polynomial of the dual of a $(\frac{2^{ms}-1}{2^s-1}, m, s, \mu, 2)$-polynomial code has α^h as a root if and only if h is divisible by 2^s-1 and

$$\max_{0 \leq \ell < s} W_{2^s}(h2^{\ell}) = j(2^s-1) \tag{54}$$

with $0 \leq j \leq \mu$.

From Equation (51) and Equation (54) we obtain

Theorem 10: The μs^{th} order binary P.G. code is a subcode of the duel of a binary $(\frac{2^{ms}-1}{2^s-1}, m, s, \mu, 2)$-polynomial code.

In reference [2], it has been shown that all the $(m-\mu-1)$-flats, ..., $(m-1)$-flats of P.G. $(m-1, 2^s)$ are contained in the $(\frac{2^{ms}-1}{2^s-1}, m, s, \mu, 2)$-polynomial code. Therefore, the dual code of a $(\frac{2^{ms}-1}{2^s-1}, m, s, \mu, 2)$-p-code is $(m-\mu-1)$-step orthogonalizable, and can correct $\lfloor \frac{J}{2} \rfloor$ errors, where

$$J = \frac{2^{(\mu+1)s}-1}{2^s-1} . \tag{55}$$

$\lfloor \frac{J}{2} \rfloor$ is also the error correcting ability of a μs^{th} order P.G. code of Weldon. Therefore, for the same designed error-correcting ability, the dual code of a polynomial code with $b = 2^s-1$ is more efficient than the corresponding P.G. code.

Projective geometry codes over GF(p) have been studied by Goethals and Delsarte [9] where p is a prime. Let $C(m-\mu-1, m-1, p^s)$ be the code generated by the set of all $(m-\mu-1)$-flats of P.G. $(m-1, p^s)$. They defined a projective geometry code as follows:

Definition (Goethals and Delsarte) [9]: A μth order projective geometry code over GF(p) of length $\frac{p^{ms}-1}{p^s-1}$ is the dual code of $C(m-\mu-1, m-1, p^s)$. Its generator polynomial is proved to have $\alpha^{t(p^s-1)}$ as a root if and only if

$$W_{p^s}[t(p^s-1)p^{\ell}] \leq \mu(p^s-1)$$

for any ℓ, $0 \leq \ell < s$, where $0 < t \leq \frac{p^{ms}-1}{p^s-1}$.

From the above definition, Theorem 1 and Theorem 2, it is obvious that the $C(m-\mu-1, m-1, p^s)$ code is identical with the $(\frac{p^{ms}-1}{p^s-1}, m, s, \mu, p)$-polynomial code, or the μth order projective geometry code of Goethals and Delsarte is the dual code of a $(\frac{p^{ms}-1}{p^s-1}$, $m, s, \mu, p)$-polynomial code.

Now, we define a <u>μth order q-ary P. G. code as the dual of a $(\frac{q^{ms}-1}{q^s-1}, m, s, \mu, q)$-polynomial code</u>. A μth order q-ary P. G. code contains all the $(m-\mu-1)$-flats of P. G. $(m-1, q^s)$ in its null space. Thus, this code is $(m-\mu-1)$-step orthogonalizable [18].

5. Conclusion

In this paper, we have studied two subclasses of polynomial codes. The relationships between these two subclasses of polynomial codes and geometry codes have been established. A BCH lower bound on the minimum distance of Euclidean geometry codes has been obtained. This bound indicates that E.G. codes are in general more powerful than Weldon's majority-logic decoding scheme has been able to demonstrate. More efficient use of E.G. codes depends on a more effective decoding scheme. A general formula for the number of parity check digits of a polynomial code studied in this paper has not yet been obtained.

Acknowledgments

Part of the actual work described herein was done by Drs. T. Kasami, W. W. Peterson and the author jointly. The author wishes to thank Dr. E. J. Weldon, Jr. of the University of Hawaii, for several stimulating discussions on geometry codes.

REFERENCES

1. Lin, S., Peterson, W. W., and Weldon, E. J., Jr., "Problems in Information Processing," Final Report, Air Force Cambridge Research Labs., Bedford, Massachusetts, (1967).

2. Kasami, T., Lin, S., and Peterson, W. W., "Polynomial Codes," Accepted for publication in IEEE Trans. on Information Theory (1968).

3. Rudolph, L. D., "Geometric Configuration and Majority Logic Decodable Codes," MEE Thesis, University of Oklahoma (1964).

4. Weldon, E. J., Jr., "Difference-Set Cyclic Codes," BSTJ, 45, 1045-1055 (1966).

5. Graham, R. L., and J. MacWilliams, "On the Number of Parity Checks in Difference-Set Cyclic Codes," BSTJ, 45, 1046-1070 (1966).

6. Kasami, T., Lin, S., and Peterson, W. W., "Some Results on Cyclic Codes Which are Invariant Under the Affine Group," Scientific Report AFCRL-66-622, Air Force Cambridge Research Labs., Bedford, Massachusetts, (1966).

7. Weldon, E. J., Jr., "Euclidean Geometry Cyclic Codes, "Proceedings of Symposium of Combinatorial Mathematics at the University of North Carolina, Chapel Hill, North Carolina, (April 1967).

8. Weldon, E. J., Jr., "Non-primitive Reed-Muller Codes," to be published IEEE Trans., IT-13, (March 1968).

9. Goethals, J. M., and Delsarte, P., "On a Class of Majority-Logic Decodable Cyclic Codes," IEEE Trans, IT-13, (March 1968).

10. Kasami, T., Lin, S., and Peterson, W. W., "New Generalizations of the Reed-Muller Codes - Part I: Primitive Codes," to be published, IEEE Trans., Vol. IT-13, (March 1968).

11. MacWilliams, F. J., and Mann, H. B., "On the p-Rank of the Design Matrix of a Difference Set," MRC Technical Summary Report No. 803, Mathematics Research Center, United States Army, University of Wisconsin, (October 1967).

12. Kasami, T., Lin, S., and Peterson, W.W., "Further Results on Generalized Reed-Muller Codes" Accepted to publish in the Journal of the Institute of Communications Engineers of Japan, (1968).

13. Berlekamp, E. R., Algebraic Coding Theory, McGraw Hill, (1968).

14. Delsarte, P., "A Geometric Approach to a Class of Cyclic Codes," Report R68, M.B.L.E., Laboratoire de Recherches, Bruxelles, Belgium, October, 1967.

15. Smith, K. J. C., "Majority Decodable Codes Derived from Finite Geometries," Institute of Statistics Mimeo Series No. 561, University of North Carolina, Chapel Hill, North Carolina, December 1967.

16. Peterson, W. W., and Weldon, E. J., Jr., Error-Correcting Codes, Second Edition, Wiley (1969).

17. Weldon, E. J., Jr., "Some Results on Majority Logic Decoding" To appear in the Proceedings of Symposium on Error Correcting Codes, Mathematics Research Center, U. S. Army, University of Wisconsin, May 1968.

18. Massey, J. L., Threshold Decoding, MIT Press, (1963).

19. Bose, R. C., and Ray, D. K. - Chaudhuri, "On a Class of Error Correcting Binary Group Codes," Information and Control, Vol. 3, pp. 68-79, (1960).

This work was supported in part by NASA Grant NGR-12-001-046.

E. J. WELDON, JR.

Some Results on Majority-Logic Decoding

1. Background

The first use of majority-logic decoding went unrecorded as such. The simple code produced by repeating a message several times has been used since antiquity and is, of course, decoded by voting. In fact, this (n, 1) code is a special case of the more modern majority-logic-decodable cyclic codes based on finite geometries.

In the recent, post-Shannon era, the first recorded use of majority-logic decoding is the work of Reed (1954) who presented an algorithm for decoding the codes discovered by Muller (1954). Prange (1958, 1959) applied similar concepts to cyclic codes and showed that several codes could be majority-logic decoded. Interestingly, the codes considered by both Muller and Prange are simply constructed in terms of finite geometries.

Reed's ideas were used by Yale (1958) and Zierler (1958) who showed that the binary maximum-length sequence codes could be decoded in this manner. Green and San Soucie (1958) showed that the binary (15, 5) cyclic code could be majority-logic decoded. All the codes examined by these investigators were actually equivalent to cyclic Reed-Muller codes -- a fact that was not known until much later.

Independently, Gallager (1960) developed ideas closely related to those of Prange and Reed; his "low-density codes" are decoded with a slightly modified version of majority-logic decoding. Gallager showed that this type of decoding could be used to attain what was promised by Shannon's noisy channel theorem.

The work of Prange on cyclic codes was carried on by Mitchell, et al. (1961). These investigators developed methods of constructing certain majority-logic-decodable cyclic codes, notably the binary Hamming and $(2^m-1, m+1)$ codes. Slightly later, Massey (1963) showed that all binary BCH codes of length 15 or less could be decoded using majority logic. More importantly, he applied this type of decoding to convolutional codes for the first time; it remains today as the only practical means of correcting random errors with these codes.

Shortly after this work, Rudolph (1964) in a Master's thesis at the University of Oklahoma, applied the notions of finite geometry to the problem of constructing majority-logic-decodable cyclic codes The result was the most important contribution to this problem since the original work of Reed and Muller. Using both Euclidean and Projective Geometries, he showed that two large classes of cyclic codes could be majority-logic decoded. Codes with useful values of n, k and d were discovered which could be decoded, in many cases, mor simply than other known codes. Furthermore, he showed that codes with symbols from GF(p), p prime, could be constructed with this method. Interestingly, all previously known majority-logic-decodabl codes, including the Reed-Muller codes, are special cases of the finite-geometry codes discovered by Rudolph.

Several major questions about these new codes remained unanswered, however. In particular, the generator polynomial, the number of information symbols and the minimum distance of the codes could only be determined in specific cases by computer.

Mathematical techniques for answering these questions were developed by Kasami, Lin, and Peterson (1966), and by Graham and MacWilliams (1966) while investigating an important subclass of the codes — Difference-Set codes — rediscovered by Weldon (1966). Their methods were used by Weldon (1967a, 1967b) to determine the minimum distance and generator polynomial of the Euclidean Geometr and Projective Geometry codes. This author also presented a method of decoding these codes which was superior to that originally devise by Rudolph. Working independently, Goethals and Delsarte (1967) derived similar results for the Projective Geometry codes. MacWilli and Mann (1967) and, independently, Smith (1967) obtained a combinatorial expression for the number of information symbols for a sub class of the finite-geometry codes.

Dasami, Lin, and Peterson (1966, 1968a) investigated the Re Muller codes and proved that these codes are equivalent to cyclic codes. They then generalized this notion to the non-binary case, di covering several important properties of cyclic codes in the process. More recently, these authors (Kasami, Lin, and Peterson (1968b)) ha defined a class of "Polynomial Codes" which includes the finite geor etry codes, the BCH codes, and the generalized Reed-Muller codes special cases.

Majority-Logic-decodable codes are attractive because their decoding algorithm is simple to implement. Now the complexity of th type of decoder increases rapidly (exponentially, in fact) with the number of levels of majority logic which must be employed or, in the terminology of Massey (1963), with the number of steps. It is clear important, therefore, to decode the finite geometry codes in as few steps as possible. It is known that the Reed-Muller and Euclidean Geometry codes can be decoded in $r + 1$ steps where r is the order of the code. Similar results hold for Projective Geometry codes (see Weldon, 1967a, 1967b).

In Part 2 of this paper, necessary conditions under which a linear code can be decoded in 1 and $L > 1$ steps. Part 3 describes a means of decoding a Euclidean Geometry code in fewer than $r + 1$ steps, where r is the order of the code. In Part 4 it is shown that a modification of majority-logic decoding can be used to decode any Euclidean or Projective Geometry code in two steps.

2. Necessary Conditions for Majority Logic Decoding

Throughout this paper cyclic (n,k) codes with symbols from $GF(q)$ are considered. The reader is assumed to be familiar with the material in Peterson (1961) on cyclic codes.

In this section conditions for 1-step and L-step decoding are established. Some additional material on majority-logic decoding is included for completeness.

The concept of orthogonal parity check sums is central to majority logic decoding. Let $s_1, s_2, \ldots, s_j$ denote check sums on various noise digits. Now if a particular error digit e_1, weighted by a $GF(q)$ coefficient a, is involved in every sum in the set, and no other error digit is checked by more than one sum, then the sums are said to be orthogonal on e_1. More generally, if every sum checks $e_1, e_2, \ldots, e_i$ with coefficients $a, b, \ldots, j$ and no other error digit appear in more than one sum, then the check sums are orthogonal on the sum $ae_1 + be_2 + \ldots + je_i$. For example, in Equation 1 s_1, s_2 and s_3 are orthogonal on the sum $ae_1 + be_2$.

$$\begin{aligned} s_1 &= ae_1 + be_2 \\ s_2 &= ae_1 + be_2 + ce_3 \\ s_3 &= ae_1 + be_2 \qquad\quad + de_4 + fe_6 \end{aligned} \tag{1}$$

Of course, if the sums are orthogonal on only one digit, it is possible to choose its coefficient to be unity, since the code forms a vector space over $GF(q)$. The following theorem is the reason for considering orthogonal check sums.

Theorem 1 (Reed-Massey). If in a linear code there are at least $d-1$ check sums orthogonal on each digit, then the code has minimum distance at least d.

Proof: It will be shown that every non-zero code word has weight at least d. Consider one of the non-zero digits in a code word. Call this the ith digit and denote the first d-1 check sums orthogonal on it by $s_{i1}, s_{i2}, \ldots, s_{i(d-1)}$. Now all these sums are zero because the word is a code word. Yet each equation has a non-zero entry so there must be at least one other non-zero digit in each sum. Because the sums are orthogonal there must be at least d-1 non-zero digits besides the ith.

Corollary 2: If it is possible to construct a set of d-1 check sums orthogonal on any digit in a cyclic code, then the code has minimum distance at least d.

Now the orthogonal check sums have another function in addition to providing a lower bound on minimum distance; they can be use to decode the code. Consider a linear code in which there are at leas d-1 check sums orthogonal on each digit and assume that $[\frac{d-1}{2}]$ or fewer errors occur. If the ith error digit is zero, then at least $d-1-[\frac{d-1}{2}]$ or at least half of the sums orthogonal on it will be zero. Because the code is linear it is always possible to choose the coefficient of the ith digit to be unity. If, on the other hand, the digit has value $v \neq 0$, then at least $d-1-([\frac{d-1}{2}]-1)$ or at least half of the sums will have value v. For the other $[\frac{d-1}{2}]-1$ errors can affect only $[\frac{d-1}{2}]-1$ or fewer of the orthogonal sums. Thus, the value of each noise digit is given by the value assumed by a clear majority of check sums orthogonal on it; if no value is assumed by a clear majori the noise digit is zero.

If the code is cyclic, then there are d-1 check sums on each digit. Furthermore, the jth sum on each error digit involves the err digits in the same relative locations. That is, if s_{ij} is the sum of the error digits at locations $\alpha^{b_1} \cdot \alpha^{b_2}, \ldots$ then $s_{(i+1)j}$ is the sum of the error digits at locations $\alpha^{b_1+1}, \alpha^{b_2+1}, \ldots$. It follows that the code can be decoded with one level of majority logic.

The first step in decoding is, as usual, the calculation of the syndrome. This is performed by an (n-k)-stage feedback shift regis ter which divides the received polynomial by g(x), the generator pol nomial of the code. The d-1 check sums orthogonal on the first err digit with coefficient unity are formed by the d-1 GF(q) adders and their scalar multipliers. The majority gate produces at its output the value assumed by the majority of its inputs, or 0 if a clear majority is assumed by no element of GF(q). This value is then subtracted from the first received information digit as the information register a syndrome generator are shifted once. It is well known that the syndrome generator, after shifting, contains the syndrome of the receiv word shifted one place to the right. Thus, the new inputs to the maj ity gate are the check sums orthogonal on the second received digit, so this digit can be corrected exactly as was the first one. Clearly a total of k-1 shifts are necessary to correct all the errors in the ir formation section of the received word. Errors in the parity section can be corrected in this manner also if desired.

This type of decoding has been referred to as 1-step decoding since only one level of majority logic is necessary.

It should be clear that each of the q^{n-k} parity check sums known to the decoder corresponds to one of the n-tuples in the null space of the code. The following theorem gives a necessary condition for 1-step decoding an (n, k) code.

Theorem 3. Let $\bar{d}$ denote the minimum distance of the dual code of an (n, k) code. Then the number of errors which can be corrected by 1-step majority logic decoding, t_1, is bounded by

$$t_1 \leq \frac{n-1}{2(\bar{d}-1)} \tag{2}$$

Proof: Because the minimum weight in the null space is $\bar{d}$, there must be at least $\bar{d}$ digits in each sum. One of these appears in all; the other $\bar{d}$-1 appear in only one of the sums. Since there are n-1 error digits in addition to the one on which the equations are orthogonal, it is possible to construct at most $(n-1)/(\bar{d}-1)$ orthogonal equations. But t_1 is at most half this number so the theorem is proved.

Q.E.D.

For many codes this result severely limits the number of errors which can be corrected with 1-step decoding. The Golay triple-error-correcting (23,12) code has $t_1 = 1$, for example. More importantly, very few errors can be corrected in this manner for most Reed-Solomon codes. For the dual code of an R-S code with distance d is an R-S with distance n-d+2, a large number. For example, a (63, 32) RS code over $GF(2^6)$ has $\bar{d} = 32$, $\bar{d} = 33$ but $t_1 = 0$. Nevertheless, there are some interesting cyclic codes for which t_1 is large compared to $[\frac{d-1}{2}]$ and these will be examined shortly. Codes for which $t_1 = [\frac{d-1}{2}]^2$ are said to be completely orthogonalizable in one step.

If a set of at least d-1 orthogonal check sums can be formed on a linear combination of error digits, s, then it can be argued that the value of s is given by the value assumed by a majority of the check sums (or zero if no element of GF(q) has a clear majority), provided $[\frac{d-1}{2}]$ or fewer errors occurred. If a number of linear combinations of noise digits can be found in this way, then together with the q^{n-k} original check sums, these form a new larger set of known check sums. (This new set corresponds to a vector space which contains the null space of the code as a proper subspace.)

Now if it is possible to carry out this procedure several times — each time obtaining at least d-1 orthogonal check sums — until a set of check sums orthogonal on a single error digit is obtained, then the value of that digit can be correctly determined. If it is possible to do this for all n error digits, then the code is said to be L-step orthogonalizable, where L is the number of levels of majority logic required. Clearly if the code is cyclic and $[\frac{d-1}{2}]$ or fewer errors occurred, correct decoding results if one digit can be decoded correctly.

This decoding procedure was first devised for the Reed-Muller codes and so is also called the Reed algorithm. Not surprisingly, L-step decoding is more powerful than 1-step.

Theorem 4. Let $\bar{d}$ denote the minimum distance of an (n,k) code. Then the number of errors which can be corrected with L-step majority logic decoding (the Reed algorithm), t_L, is bounded by

$$
\begin{aligned}
t_L &\leq \frac{n}{\bar{d}} - \frac{1}{2} \quad ; \ \bar{d} \text{ even} \\
&\leq \frac{n+1}{\bar{d}+1} - \frac{1}{2} ; \ \bar{d} \text{ odd}
\end{aligned}
\tag{3}
$$

Proof: In order to correct t_L errors it must be possible to construct at least $2t_L$ check sums orthogonal on a set of digits common to each sum. Call the number of bits in this set B. Clearly, $B \leq [\frac{\bar{d}}{2}]$. For otherwise the sum of the vectors corresponding to two sums orthogonal on the set of B digits, which is also in the null space of the code, would have weight less than $\bar{d}$.

Since the $2t_L$ equations have at most $[\frac{\bar{d}}{2}]$ digits in common and at least $\bar{d}$ digits in all, it must be that

$$2t_L(\bar{d} - [\frac{\bar{d}}{2}]) \leq n - [\frac{\bar{d}}{2}] \ .$$

The theorem follows. Q.E.D.

This result states that, for a given code, one can hope to correct roughly twice as many errors with L-step decoding as with 1-step. It also shows that some codes cannot be majority-logic decoded up to their error-correcting ability by either 1 or L-step decoding. The Golay code and most other quadratic residue codes with known d are of this type, as are the Reed-Solomon codes (with a few trivial exceptions). Some, but not all, BCH codes for which d is known also have $d\bar{d} > 2n$. An attempt has been made to determine under what conditions this does or does not occur; so far it has proved unsuccessful.

3. An Improvement on the Reed Decoding Algorithm for Euclidean Geometry Codes

The Reed algorithm applies to the original Reed-Muller (RM) codes and to their two generalizations, the Projective Geometry (PG) and Euclidean Geometry (EG) codes. The improvement described herein is applicable only to the RM and EG codes.

For a prime p, an augmented cyclic p-ary Euclidean Geometry code of order r has the following properties:

$$n = p^{ms}$$

$$\text{associated geometry} = EG(m, p^s)$$

and the additional property that the polynomial associated with every (r+1)-flat in this geometry is in the null space of the code. Now given all the (r+1)-flats one determines the r-flats by means of the Reed algorithm. Since there are $p^{s(r+1)}$ points in an (r+1)-flat, there are

$$J = \frac{p^{sm} - p^{sr}}{p^{q(r+1)} - p^{sr}} = \frac{p^{s(m-r)} - 1}{p^s - 1}$$

$$= p^{s(m-r-1)} + p^{s(m-r-2)} + \dots + p^s + 1$$

(r+1)-flats which intersect only on a particular r-flat. Thus, if $t = [\frac{J}{2}]$ or fewer errors occur, it is possible to determine the check sum associated with the r-flat by simply taking the majority vote of the check sums of the (r+1)-flats intersecting on it. This can be done for every r-flat in the geometry. Then the (r-1)-flats can be determined in a similar way from the r-flats. Finally, the 0-flats (noise digits) can be determined and error correction performed by adding these digits to the corresponding data symbols. In Massey's terminology we can say that the code can be orthogonalized in r+1 steps.

When the integer ms is a composite number, it is possible to reduce the number of steps considerably. Since the cost of a Reed decoder increases very rapidly with L, the number of steps, this can effect a substantial savings in equipment.

The improved algorithm operates as follows. The first step is identical to the original algorithm. That is, the r-flats are determined by a majority decision of the intersecting (r-1)-flats. Now there are two cases to consider. If $(r, m) = 1$, then the second step is also identical to that of the original algorithm, i.e., the (r-1)-flats are determined from the r-flats. We proceed until the c-flats are determined where $(c, m) = f \neq 1$. Let $c/f = c'$ and let $m/f = m'$. Now a c'-flat in $EG(m', p^{sf})$ is also an fc'-flat in $EG(fm', p^s)$. (The converse is not necessarily true, however.) Thus, we can now consider the geometry $EG(m', p^{sf})$ which has considerably lower dimension than the original. Since every fc'-flat in $EG(fm', p^s)$ is known, so is every c'-flat in $EG(m', p^{sf})$. The number of c'-flats orthogonal on a (c' - 1)-flat is

$$J' = \frac{p^{sfm'} - p^{sf(c'-1)}}{p^{sfc'} - p^{sf(c'-1)}} = p^{sf(m'-c')} + p^{sf(m'-c'-1)} + \ldots + p^{sf} + 1 .$$

Since $r+1 > c$, it is not difficult to verify that $J' \geq J$ for all choices of p, r, c, s, and m . The details are omitted here. The next step in the new algorithm is to determine the (c'-1)-flats knowing the c'-flats. Because $J' \geq J$ this is possible even in the presence of $t = [\frac{J}{2}]$ or fewer errors. Therefore, the new algorithm does not decrease the error-correcting capability of the decoder.

Given the (c' - 1)-flats it is possible to determine the (c'-2)-flats, etc. Alternately one can apply the same trick again and so reduce the number of steps still further.

An example is called for. Let $p = 2$, $ms = 12$, and consider the order-10 Reed-Muller code. The associated geometry is $EG(12, 2)$ and all 11-flats of the geometry are in the null space of the code. Given the 11-flats, we determine the 10-flats in the ordinary way. These can be regarded as 5-flats in $EG(6, 2^2)$ since $f = (10, 12) = 2$. Thus, in the next step we determine the 4-flats of $EG(6, 2^2)$ (or 8-flats of the original geometry).

Now again the new c and m are not relatively prime, $(4, 6) = 2$, so we can regard the 4-flats of $EG(6, 2^2)$ as 2-flats of $EG(3, 2^4)$. Then the third step in decoding consists of determining the 1-flats of $EG(3, 2^4)$ (or the 4-flats of the original geometry. The fourth and last step is to determine the 0-flats of $EG(3, 2^4)$ from the 1-flats. This procedure is depicted in the following chart:

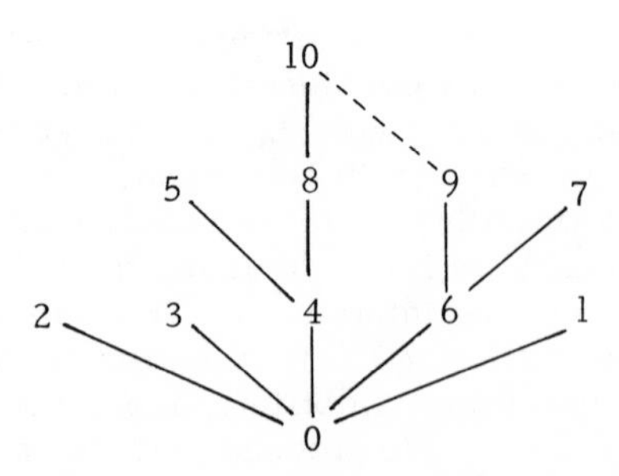

Decoding Chart for EG and RM Codes of length 4096

The numbers in the chart represent the dimensions of the various flats in $EG(12, 2)$. The decoding path just described begins with the 11-flats of EG(12, 2), proceeds to the 10-flats, to the 8-flats, 4-flats, and finally 0-flats. Thus, the original 11-step decoding has been reduced to 4 steps. Since the first step in decoding any EG or RM code is always specified, it is not shown in the chart. The other paths represent the decoding procedures for other RM and EG codes. For example, the order-7 RM code, which has 8-flats in its null space, has the path 7 - 6 - 0 .

As another example, let $p = 2$, $m = 12$ and consider the order-4 EG code associated with $EG(6, 2^2)$. This code has all 5-flats in its null space. The first step in decoding is to determine the 4-flats of $EG(6, 2^2)$ from the 5-flats. This done, we apply our new algorithm. Since the known flats are 8-flats in EG(12, 2), from the decoding chart the shortest path is 8 - 4 - 0 . Thus, the original 5-step decoding process is reduced to 3 steps.

The dotted path (from 10 to 9) shown in the chart represents an alternate, equally short path for the order-10 RM code. In general, such paths may be shorter than the one presented above (whose major merit is that it can be described simply). Obviously, the best decoding procedure can be found by constructing the decoding chart and then choosing the shortest path. The charts for all codes with $m \leq 16$ are shown in Figure 1. Charts for the case of ms prime are omitted since they consist of a single path from $m - 2$ to 0 .

In conclusion, the following remarks are pertinent. The proposed improvement is a real one. If one had to build a Reed decoder for a long code, the savings effected by using the modified algorithm would be enormous in most cases. However, the codes to which the algorithm applies deteriorate rather rapidly as their length increases. Thus, the main application of this type of decoding appears to be to relatively short codes, i.e., codes of up to say, 500 bits in length. In this range the savings are much less, although still substantial in some cases.

4. A Modification of the Reed Algorithm Capable of Decoding All EG and PG Codes in Two Steps

An order-r PG code has all r-flats of $PG(m, p^s)$ in its null space. As seen in Section 3, for these codes

$$J = \frac{p^{s(m-r+1)} - 1}{p^s - 1} \tag{4}$$

Now the first step in the modified decoding procedure will be to determine all r-flats — exactly as before. If $[\frac{J}{2}]$ or fewer errors occurred, all these flats are determined correctly.

For codes with order one this completes the decoding process. Codes with order two can be decoded with one additional step using the standard majority logic algorithm. For all other codes, however, at least two more steps are required; this is costly since complexity grows exponentially with the number of steps.

For these codes the following method can be used. At step 1 determine all (r-1)-flats which intersect the high-order digit of the code. In step 2 the 0-flats are determined from these flats.

Let M denote the number of (r-1)-flats in $PG(m,p^s)$. From Mann (1949)

$$M = \frac{(p^{sm}+p^{s(m-1)}+\dots+p^s+1)(p^{sm}+p^{s(m-1)}+\dots+p^s)\dots(p^{sm}+\dots+p^{s}}{(p^{s(r-1)}+p^{s(r-2)}+\dots+p^s+1)\dots(p^{s(r-1)}+p^{s(r-2)})p^{s(r-1)}}$$

Now each (r-1)-flat contains $p^{s(r-1)}+p^{s(r-2)}+\dots+p^s+1$ points. Hence, the number of (r-1)-flats passing through a given point is

$$N = \frac{M(p^{s(r-1)}+p^{s(r-2)}+\dots+p^s+1)}{p^{sm}+p^{s(m-1)}+\dots+p^s+1}$$

$$= \frac{(p^{sm}+p^{s(m-1)}+\dots+p^s)\dots(p^{sm}+\dots+p^{s(r-1)})}{(p^{s(r-1)}+p^{s(r-2)}+\dots+p^s)\dots(p^{s(r-1)}+p^{s(r-2)})p^{s(r-1)}}.$$

Also, each (r-1)-flat contains

$$\frac{(p^{s(r-1)}+\dots+p^s+1)(p^{s(r-2)}+\dots+p^s+1)}{(p^s+1)}.$$

lines or 1-flats. Hence, the number of (r-1)-flats passing through a given line is

$$\lambda = \frac{M(p^{s(r-1)}+\dots+p^s+1)(p^{s(r-2)}+\dots+p^s+1)/(p^s+1)}{(p^{sm}+\dots+p^s+1)(p^{s(m-1)}+\dots+p^s+1)/(p^s+1)}$$

where the denominator is just the number of 1-flats in $PG(m,p^s)$.

Consider the N check sums on a particular digit corresponding to the (r-1)-flats passing through a point. Each other digit appears in λ of these sums since the two points define a line and there are λ (r-1)-flats on each line.

Note that

$$\frac{N}{\lambda} = \frac{p^{s(m-1)} + \ldots + p^{s} + 1}{p^{s(r-2)} + \ldots + p^{s} + 1} = \frac{p^{sm} - 1}{p^{s(r-1)} - 1} \quad .$$

If $[\frac{N}{2\lambda}]$ or fewer errors occurred, then the value of the "central" error digit must be correctly given by the value assumed by the majority of the $(r-1)$-flats, or 0 in the case of a tie. But for $m > r > 1$, it is easily proved that $\frac{N}{\lambda} > J$; therefore, correct decoding results if $[\frac{J}{2}]$ or fewer errors occurred.

Identical results can be proved for EG codes. Thus, in cases where these codes cannot be decoded in one or two steps with the improvement described in Section 3, they can be decoded in two steps using this approach.

This idea of using non-orthogonal check sums is due to Rudolph (1964). He showed that this approach could be used to correct a substantial number of errors in a finite geometry code in one step. It is not possible to correct all error patterns of weight $[\frac{J}{2}]$ in this manner, however.

In closing, it should be pointed out that while any finite geometry code can be decoded in two steps using this modified algorithm, it may be more economical to use the conventional algorithm, possibly with the improvement described in Section 3, even if more than two steps are involved. The reason for this is that N in Equation 5 can be <u>very</u> large number even for reasonable values of p, m, and s, and an N-input majority gate is required with the modified algorithm.

Acknowledgment

The author would like to thank Dr. Shu Lin for many helpful discussions on the material presented in this paper.

REFERENCES

1. Reed, I. S., "A Class of Multiple-Error-Correcting Codes and the Decoding Scheme," IRE Trans., IT-4, 38-49 (1954).

2. Muller, D. E., "Applications of Boolean Algebra to Switching Circuit Design and to Error Detection," IRE Trans., EC-3, 6-12 (1954).

3. Prange, E., Some Cyclic Error Correcting Codes with Simple Decoding Algorithms, AFCRC-TN-38-156, Air Force Cambridge Research Center, Cambridge, Mass. (1958).

4. Prange, E., The Use of Coset Equivalence in the Analysis and Design of Group Codes, AFCRC-TR-59-164, Air Force Cambridge Research Center, Cambridge, Mass. (1959).

5. Yale, R. B., Error Correcting Codes and Linear Recurring Sequences, Lincoln Laboratory Report 34-77, Lincoln Labs, MIT (1958).

6. Zierler, N., On A Variation of the First Order Reed-Muller Codes, Lincoln Laboratory Report 34-80, Lincoln Labs, MIT (1958).

7. Green, J. H., and San Soucie, R. L., "An Error-Correcting Encoder and Decoder of High Efficiency," Proc. IRE, 46, 1741-1744 (1958).

8. Gallager, R. G., "Low-Density Parity Check Codes," IRE Trans., IT-8, p. 21 (1962). Also ScD. Thesis, MIT (1960).

9. Mitchell, M. E., et al., Coding and Decoding Operations Research, Final Report on Contract AF 19 (604)-6183, AFCRL8, 1961.

10. Massey, J. L., Threshold Decoding, MIT Press (1963).

11. Rudolph, L. D., "Geometric Configureation and Majority Logic Decodable Codes," MEE Thesis, University of Oklahoma, Norman, Oklahoma (1964).

12. Kasami, T., S. Lin and W. W. Peterson, "Some Results on Cyclic Codes Which Are Invariant Under the Affine Group," Scientific Report, AFCRL 66-622, Air Force Cambridge Research Laboratory, Bedford, Mass. (1966).

13. Graham, R. L., and J. MacWilliams, "On the Number of Parity Checks in Difference-Set Cyclic Codes," BSTJ, 45, 1046-1070 (1966).

14. Weldon, E. J., Jr., "Difference-Set Cyclic Codes," BSTJ, 45, 1045-1055 (1966).

15. Weldon, E. J., Jr., "Euclidean Geometry Cyclic Codes," Proceedings of Symposium of Combinatorial Mathematics at the University of North Carolina, Chapel Hill, North Carolina (April, 1967a).

16. Weldon, E. J., Jr., "Non-Primitive Reed-Muller Codes," AFCRL-67-0177, Scientific Report No. 12 (1967b). To appear, IEEE Trans., IT-14 (1968).

17. Goethals, J. M., and Delsarte, P., "On a Class of Majority-Logic Decodable Cyclic Codes," IEEE Trans., IT-14 (1968).

18. MacWilliams, F. J., and Mann, H. B., "On the p-Rank of the Design Matrix of a Difference Set," MRC Technical Summary Report No. 803, Mathematics Research Center, United States Army, University of Wisconsin (1967).

19. Smith, K. J. C., "Majority Decodable Codes Derived from Finite Geometries," Institute of Statistics Mimeo Series No. 561, University of North Carolina, Chapel Hill, North Carolina (1967).

20. Kasami, T., Lin, S., and Peterson, W. W., "New Generalizations of the Reed-Muller Codes — Part I: Primitive Codes," to be published, IEEE Trans., Vol. IT-14 (1968a).

21. Kasami, T., Lin, S., and Peterson, W. W., "Polynomial Codes," to appear, IEEE Trans., IT-14 (1968b).

22. Peterson, W. W., Error-Correcting Codes, Wiley, New York (1961).

23. Mann, H. B., Analysis and Design of Experiments, Dover Press (1949).

This research was supported by the
National Aeronautics and Space Administration
Grant NGR-12-001-046

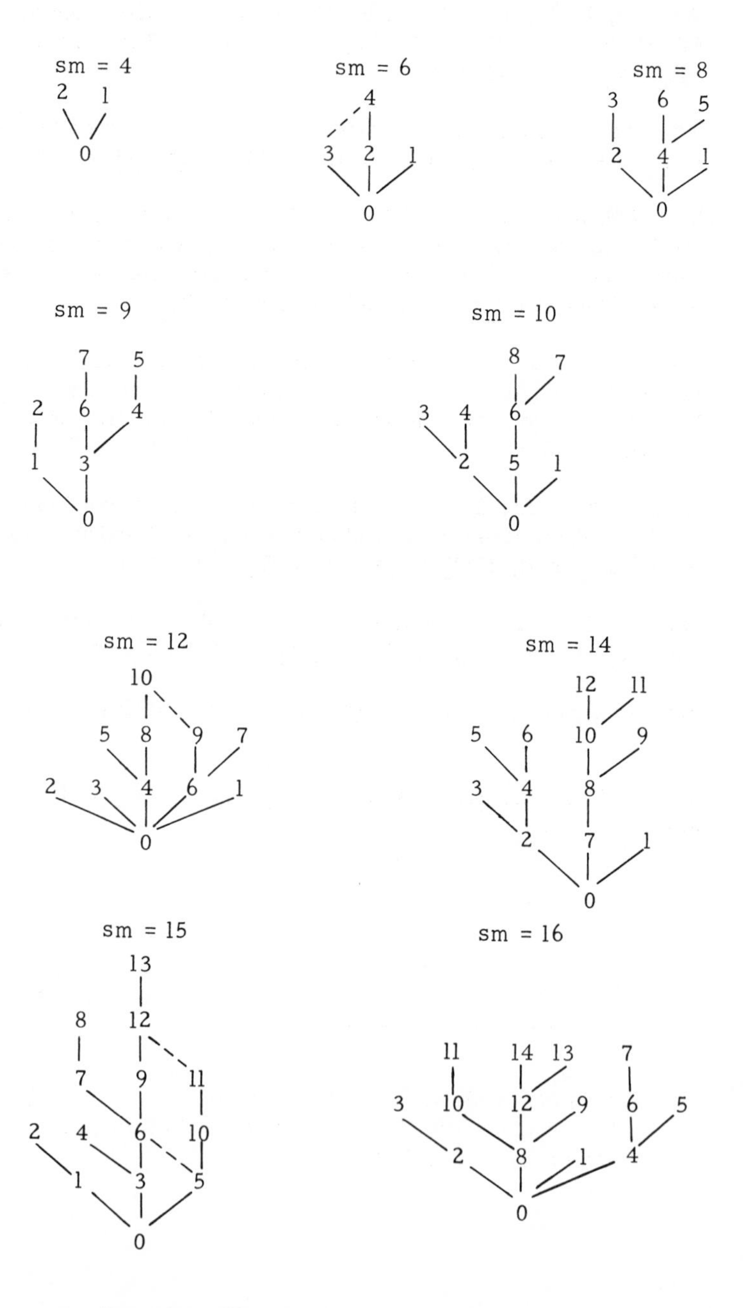

Figure 1. Decoding Charts of EG and RM codes for sm ≤ 16, sm not prime.

P. CAMION

Unimodular Modules and Cyclotomic Polynomials

Abstract

This work is concerned with results of L. Redei, H. Mann, N. G. De Bruijn and A. Sands relative to cyclotomic polynomials and was suggested by M. Schützenberger. It actually appears that the properties proved by those authors are strongly related to some of our own results on unimodular modules. The theorem of Redei-De Bruijn, proved independently by Schoenberg, asserts that the ideal $F_n(X)\underline{Z}[X]$ ($F_n(X)$ being the cyclotomic polynomial with degree $\Phi(n)$) is also the ideal spanned by the polynomials $F_p(X^{n/p})$ where p runs over all prime factors of n. This is immediately proved by a recurrent use of the following lemma: $F_{p_1 p_2 \cdots p_{k-1}}(X)$ and $F_{p_k p_2 \cdots p_{k-1}}(X)$, where the p_i are primes, verify a Bezout relation in $Z[X]$. We prove the following generalization of this lemma. Let $(n_1, \ldots, n_r) = 1$ and $(n_i, t) = 1$, for all i. Then the polynomials $F_{n_i t}(X)$, $i = 1, \ldots, r$, verify a Bezout relation in $\underline{Z}[X]$. This generalization is a direct consequence of unimodular properties.

We also prove that for polynomials $f(X)$ which are certain well defined products of cyclotomic polynomials, the ideal generated by $f(X)$ in the algebra modulo $X^n - 1$ is (considered as a $\underline{Z}$-module) unimodular. That is especially true for $f(X) = F_{pq}(X)$, p, q primes, which is a result of H. Mann. Also for $f(X) = F_{pq}(X)$, a theorem of De Bruijn is a consequence of that property and for more general $f(X)$, a theorem of A. Sands, which generalized that one of De Bruijn and allowed him to prove the Hajos conjecture on the factorization of abelian groups.

1. Definition and General Properties

We give the definition of a Unimodular Module in the simple case we have to deal with. That is the case where the module has a finite basis. For a general definition, see [1].

Let S be a commutative ring with unity and Y be a finite set with m elements. S^Y will denote the S-module of m-tuples with coordinates in S.

The support $s(x)$ of an element $x \in S^Y$ will be the set of $i \in Y$ such that the ith coordinate x_i of x is not zero.

Now let M be a sub-module of S^Y. $\mathcal{S}(M)$ will denote the family of non empty supports of the elements of M. Let us then consider a support $s(x)$, $x \in M$, which is minimal in $\mathcal{S}(M)$. For every couple $z, y \in M$ with $s(z) = s(y) = s(x)$, one has

$$s(z-y) = s(x) \quad . \tag{1}$$

And for every $\alpha \in S$, one has

$$s(\alpha z) = \emptyset \quad \text{or} \quad s(\alpha z) = s(x) \quad .$$

Hence, 0 and the set of z with $s(z) = s(x)$ is a submodule of M which we call a minimal submodule of M. Then we have the

Definition 1. A unimodular module $M \subset S^Y$ is a submodule of S^Y in which every minimal submodule has the form $\{\alpha u / \alpha \in S\}$ where the non-zero coordinates of $u \in M$ are invertible.

Such an element $u \in M$ in which the non-zero coordinates are invertible and with minimal support is called a generator of the unimodular module M. It is a monic generator if its first non-zero coordinate is 1. It is clear, from the definition, that a unimodular module is spanned by its monic generators, the set of which is finite.

Definition 2. A matrix with entries in S is totally unimodular over S if every non-zero subdeterminant is an invertible element of S.

Definition 3. A n by m matrix with entries in S is locally unimodular over S if every n by n non-singular submatrix has a determinant invertible in S.

Definition 4. A n by m matrix is called echelon if it has the unit matrix of order n as a submatrix.

Property 1. A locally unimodular echelon matrix is totally unimodular.

Property 1 is an immediate consequence of the definition.

The following properties of a submodule M of S^Y are equivalent (the proof will be found in [1]).

1. M is a unimodular module.
2. M is spanned by the rows of an echelon matrix, locally unimodular.
3. M is spanned by the rows of a totally unimodular matrix.
4. M is spanned by the rows of a locally unimodular matrix.

As an immediate consequence of the equivalence of 1 and 2 one has the

Theorem 1. The orthogonal module of a unimodular module is unimodular.

2. Integer unimodular modules

2.1. The homomorphism of $\underline{Z}^m$ onto F_2^m restricted to unimodular modules:

F_2 is the Galois field with two elements. For generators with disjoint supports, we shall say briefly disjoint generators. Let us recall the following theorem [2].

Theorem 1. Let h denote the canonical homomorphism of $\underline{Z}^m$ onto F_2^m. A submodule M of Z^m is unimodular if and only if the inverse image of every element in $h(M)$ contains a vector which is a sum of disjoint generators and the sets of minimal supports $\mathcal{S}(M)$ and $\mathcal{S}(h(M))$ are the same.

1. Necessity of the condition:

$h(M)$ being a linear subspace of F_2^m is a unimodular module, and every one of its vectors is a sum of disjoint generators. Then we have only to consider the inverse image of a generator of $h(M)$. By Property 1.2, M being spanned by the rows of a totally unimodular echelon matrix is also the module of vectors which are orthogonal to the rows of a n by m totally unimodular echelon matrix B with rank n. Then

$$M = \{x/Bx = 0, \quad x \in \underline{Z}^m\} .$$

But $h(M)$ has dimension $m-n$, and $\bar{B}$, the homomorphic image of B modulo 2 has rank n, then the linear space

$$\{\bar{x}/\bar{B}\bar{x} = \bar{0}, \quad \bar{x} \in F_2^m\}$$

which contains $h(M)$ and has dimension $m-n$ is $h(M)$.

Now, let us consider a generator $\bar{x}$ of $h(M)$, one has

$$\sum_{j \in s(\bar{x})-\{k\}} \bar{B}_j = \bar{B}_k ,$$

where the column vectors set $\{B_j\}_{j \in s(\bar{x})}$ has rank $|s(\bar{x})| - 1$, B being totally unimodular. Thus there exist a set of $\varepsilon_j \in \{1, -1\}$ such that

$$\sum_{j \in s(\bar{x})-\{k\}} \varepsilon_j B_j = \varepsilon_k B_k ,$$

and those ε_j are clearly the non-zero components of a generator of M. The direct image of a generator of M is then clearly a generator of $h(M)$, thus $\mathcal{S}(M)$ and $\mathcal{S}(h(M))$ are identical.

2. *Sufficiency of the condition*:

We have to prove that every vector of M with minimal support has the form αu, where $\alpha \in \underline{Z}$ and u is a generator belonging to M . Let x be a vector of M with minimal support. This support being a support of $\mathcal{S}(h(M))$, we may first assume that x ha odd non-zero coordinates, because $s(x) \in \mathcal{S}(h(M))$. If u is the generator of $h^{-1}(\bar{x})$ given by the hypothesis, and α the least positive component of x or -x, let us say of x , $x - \alpha u$ is 0 because it cannot have the same support as x, hence $x = \alpha u$. Now if y has all even coordinates with $s(y) = s(x)$, $y - u$ has all odd coordinates and $y = (\alpha + 1)u$.

Corollary. *If* $\{B_j\}_{j \in J}$ *is a set of columns (rows) of a totally unimodular matrix with* $\sum_{j \in J} B_j \equiv 0 \bmod 2$, *there exist* $\varepsilon_j \in \{1, -1\}$ *such that* $\sum_{j \in J} \varepsilon_j B_j = 0$.

This follows from the fact that the module of vectors orthogon to the rows of a totally unimodular matrix is unimodular, being the or thogonal module of a unimodular module (Theorem 1.1 and Property 1.

Definition. *A Dantzig matrix or D-matrix* [3] is an n by m matrix with entries in $\underline{Z}$ with rank n where all the square sub-matrices of order n which are nonsingular have the same determinant.

Theorem 2. *Let* B *be an* n *by* m *matrix with entries in* $\underline{Z}$, $n < m$ *with rank* n . B *is a Dantzig matrix if and only if the submodule of* $\underline{Z}^m$ *orthogonal to the rows of* B *is unimodular.*

An analogous theorem has been proved by Heller [3].

1. *Sufficiency of the condition*:

Let M be the module orthogonal to the rows of B and $x \in M$ a vector with minimal support $s(x)$. $s(x) = J$ is a minimal subset of linearly dependent columns of B . Then for any $k \in J$ $\{B_j\}_{j \in J - \{k\}}$ is a free set of vectors which may be completed to a bas $\{B_j\}_{j \in J' \supset J}$ of $\underline{Q}^n$. We may express B_k by means of this basis wi coefficients in the field $\underline{Q}$ of rational numbers. We then obtain

$$\sum_{j \in J'} \alpha_j B_j = \alpha_k B_k$$

where the α_i are integers. But we have

$$\sum_{j \in J} x_j B_j = -x_k B_k \ .$$

Let $g.c.d\{\alpha_i\} = 1$ and let us first assume $g.c.d\{x_i\} = 1$. One must have $x_i = \alpha_i$, $\forall i \in J'$.

Now B being a Dantzig matrix $\alpha_i \in \{0, 1, -1\}, \forall i$, then, generally, $x = \beta u$, $\beta \in \underline{Z}$ and u a generator belonging to M which is thus unimodular.

2. Necessity of the condition:

Let B_J and B_K be two non-singular submatrices of B of order n corresponding respectively to the sets J and K of columns $\{B_j\}_{j \in J} \cup \{B_k\}$ is a set of linearly dependent columns of B, for every $k \in K - J$. Then there exist a minimal set of columns of this set containing B_k which are linearly dependent. To this set corresponds a generator of M, which means that for every $k \in K$ there exist a vector y^k such that

$$B_J y^k = B_k \tag{1}$$

and $y_i^k \in \{0, 1, -1\}, \forall i$.

Thus there exist a matrix U with integer entries such that

$$B_J U = B_K \ . \tag{2}$$

But for the same reason, there exist a matrix V with $B_K V = B_J$, and $B_J UV = B_J$, which means that $UV = I$, $\mathrm{Det}(U) = \pm 1$ and by (2) $\mathrm{Det}(B_J) = \mathrm{Det}(B_K)$.

Definition. An odd D-matrix is an $n \times m$ D-matrix in which a submatrix of order n has an odd determinant.

Theorem 3. Let B be an n by m matrix with rank n . Let M be the module orthogonal to the rows of B and $\bar{M}$ the linear space orthogonal to the rows of $\bar{B}$. Then B is an odd Dantzig matrix if and only if to each $\bar{x}$ in $\bar{M}$ corresponds a y in M with $y_i \in \{0, 1, -1\}$, $\forall i$, and $\bar{y} = \bar{x}$.

1. Necessity of the condition:

$Bx = 0 \Rightarrow \bar{B}\bar{x} = 0$, thus $h(M) \subset \bar{M}$. On the other hand, the rank of $\bar{B}$ is n and then $\bar{M}$ has dimension $m - n$. But M being unimodular by theorem 2, $h(M)$ has dimension $m - n$ and $\bar{M} = h(M)$. The thesis is thus a consequence of theorem 1.

2. Sufficiency of the condition:

One has also $h(M) \subset \bar{M}$ and the hypothesis now expresses that $\bar{M} \subset h(M)$. Now it is clear that the inverse image of every $\bar{x}$ in $h(M)$ contains a vector which is a sum of disjoint generators. Thus the thesis will be a consequence of theorems 1 and 2 if we prove that $\mathcal{S}(M) = \mathcal{S}(h(M))$.

If $x \in M$ has minimal support $s(x)$, there exist a $y \in M$ with $s(y) = s(x)$, y having at least one odd component. Thus

$\phi \neq s(\bar{y}) \subset s(y)$. Thus there exist a generator $\bar{z}$ of $h(M)$ with $s(\bar{z}) \subset s(y)$. And there exist a $u \in M$ with $s(u) = s(\bar{z})$. But y being minimal in M, $s(\bar{z}) = s(u) \subset s(y)$, whence $s(\bar{z}) = s(y)$.

On the other hand, let $\bar{x}$ be an element of $h(M)$ with minima support. There exist a $u \in h^{-1}(\bar{x})$ with $s(u) = s(\bar{x})$, by hypothesis Suppose u does not have minimal support; then there should exist a y with at least one odd coordinate and $\phi \neq s(\bar{y}) \subset\subset s(u) = s(\bar{x})$, which is impossible. This completes the proof of theorem 3.

3. Some Properties of Cyclotomic Polynomials Deduced from Propertie of Unimodular Modules and Totally Unimodular Matrices

A threorem of Redei-De Bruijn [4] also proved by Schoenberg [and differently by H. Mann [6] is an easy consequence of the totally unimodular property of "interval matrices".

The theorem may be stated as follows:

Theorem 1. The ideal of $\underline{Z}[X]$ generated by $\{F_p(X^{n/p})\}$ where p runs over the set of prime divisors of n is the principal ideal generated by the cyclotomic polynomial $F_n(X)$.

It is clear that $F_n(X)$ is a factor of $F_p(X^{n/p})$ for every prim divisor p of n . Thus it has to be proved that there exist polynomia $E(X,p) \in \underline{Z}[X]$ satisfying

$$F_n(X) = \sum_{p|n} E(X,p)F_p(X^{n/p}) \quad . \tag{1}$$

Let $v = \prod_{p|n} p$ and $n = m.v$. Since $F_n(X) = F_v(X^m)$, it is enough to prove (1) when n is squarefree. The proof is given by induction on the number of prime factors of n . One has:

$$F_{p_1 \cdots p_k}(X) \cdot F_{p_2 \cdots p_k}(X) = F_{p_2 \cdots p_k}(X^{p_1}) = \sum_{p|n/p_1} E_1(X^{p_1},p)F_p(X^{n/p}) \tag{2}$$

and

$$F_{p_1 \cdots p_k}(X) \cdot F_{p_1 \cdots p_{k-1}}(X) = F_{p_1 \cdots p_{k-1}}(X^{p_k}) = \sum_{p|n/p_k} E_k(X^{p_k},p)F_p(X^{n/p}) \quad . \tag{3}$$

The theorem will be proved by showing that $F_{p_2 \ldots p_k}(X)$ and $F_{p_1 \ldots p_{k-1}}(X)$ generate $\underline{Z}[X]$. We shall prove in fact the more general theorem that follows:

Theorem 2. *Let* $1 < n_1 < \ldots < n_\ell$ *be positive integers with* g.c.d = 1 *and* t *such that* $(n_i, t) = 1$, $i = 1, \ldots, \ell$. *Then there exist polynomials* $E_1(X,t), \ldots, E_\ell(X,t)$ *with integer coefficients such that*

$$\sum_{i=1}^{\ell} E_i(X,t) F_{tn_i}(X) = 1 \quad . \tag{4}$$

We shall first prove the theorem for $t = 1$. Let $u_{n_i}(X)$ be the polynomial $\sum_{j<n_i} X^j$. The factors of $u_{n_i}(X)$ are all the polynomials $F_d(X)$ where d is a divisor of n_i. As $(n_1, \ldots, n_\ell) = 1$, the polynomials $u_{n_i}(X)$ have no factor in common even with rational coefficients. Thus there exist polynomials $E_1'(X,1), \ldots, E_\ell'(X,1)$ with rational coefficients such that

$$\sum_{i=1}^{\ell} E_i'(X,1) u_{n_i}(X) = 1 \quad . \tag{5}$$

But with any polynomial $\sum_j a_j X^j$ we may define a row (a_j) of a matrix. Thus to the set of polynomials $\bigcup_{1 \le i \le \ell} \{X^j u_{n_i}(X)\}_{j \in \underline{N}}$ corresponds a matrix which is a well known totally unimodular matrix. (This can easily be obtained from [7]). Now (5) asserts that there exists a linear combination of the rows of that matrix with rational coefficients equal to some m-tuple with the form $(1, 0, \ldots, 0)$. Thus every subdeterminant of that matrix being 1, -1 or 0, there exist polynomials $E_j''(X,1)$ with coefficients 1, -1 or 0 such that

$$\sum_{i=1}^{\ell} E_i''(X,1) u_{n_i}(X) = 1 \quad . \tag{6}$$

Now, $F_{n_i}(X)$ is a factor of $u_{n_i}(X)$, for all i. Hence (4) is verified for $t = 1$.

It is sufficient to prove (4) for squarefree t. The argument is by induction on the number of prime factors of t. Assume (4) is proved for $t = p_1 \ldots p_{k-1}$. One has for $t = p_1 p_2 \ldots p_k$:

$$F_{n_i t}(X) F_{n_i t/p_k}(X) = F_{n_i t/p_k}(X^{p_k}), \quad i = 1, \ldots, \ell \quad . \tag{7}$$

Consequently

$$\sum_{i=1}^{\ell} F_{n_i t} \cdot F_{n_i t/p_k}(X)\, E_i(X^{p_k}, t/p_k) = 1 \ , \tag{8}$$

which proves the theorem.

<u>Corollary 1.</u> <u>Let</u> n_1 <u>and</u> n_2 <u>be squarefree positive integers with</u> $(n_1, n_2) \neq n_1, n_2$. <u>Then there exist polynomials</u> $E_1(X), E_2(X) \in \underline{Z}[X]$ <u>with the property</u>

$$E_1(X)\, F_{n_1}(X) + E_2(X)\, F_{n_2}(X) = 1 \ .$$

One takes $t = \text{g.c.d}(n_1, n_2)$. One has $n_1/t \neq 1$ and $n_2/t \neq 1$, thus theorem 2 may be applied.

<u>Corollary 2.</u> <u>Let</u> $u_k(X) = \sum_{j<k} X^j$ <u>and</u> $u_t(X) = \sum_{j<k} X^j$ <u>with</u> $k < t$. <u>If</u> P_t <u>is the companion matrix of</u> $X^t - 1$, <u>then</u> $\text{Det}(u_k(P_t)) = \pm k$ or 0 as $(k, t) = 1$ or $(k, t) \neq 1$.

Let us first have $(k, t) = 1$.

As we have seen in the proof of theorem 2, there exist polynomials $E_k(X)$ and $E_t(X)$ with coefficient 1, -1, and 0 satisfying

$$E_k(X)\, u_k(X) + E_t(X)\, u_t(X) = 1 \ . \tag{9}$$

Thus $u_k(\alpha^i)$ is invertible in $\underline{Z}[\alpha]$ for all $0 < i < t$, where α is a primitive t^{th} root of unity. Now $\text{Det}(u_k(P_t)) = \prod_{i<t} u_k(\alpha^i) = k \prod_{0<i<t} u_k(\alpha^i) \in \underline{Z}$. But a product of invertible elements of the ring $\underline{Z}[\alpha]$ is an invertible element, and if it belongs to $\underline{Z}$, it must be 1 or -1 .

On the other hand if $(k, t) \neq 1$, $u_k(X)$ and $u_t(X)$ have a common factor, thus there exist a i for which $u_k(\alpha^i) = 0$, which proves the corollary.

<u>Corollary 3.</u> <u>With the same notations as in corollary</u> 2, <u>if</u> $(k, t) = 1$, <u>the matrix</u> $F_k(P_t)$ <u>is unimodular.</u>

The proof is the same as for corollary 2.

We follow by a simple remark concerning cyclotomic polynomials and unimodular modules. Let us recall the following theorem [1].

<u>Theorem 3.</u> <u>The non-zero coordinates of a generator of a unimodular module over a ring</u> S <u>in a totally unimodular basis are units of</u> S .

and also

<u>Theorem 4.</u> <u>Let</u> M <u>be a unimodular module over a linear ordered ring</u> S . <u>Then every element of</u> M <u>with nonnegative components is a positive linear combination of generators of</u> M <u>with nonnegative components.</u>

Now, the matrix corresponding to the set of polynomials $\{X^j F_p(X)\}_{j\in \underline{N}}$ being totally unimodular, to verify the well-known property that $F_{pq}(X)$ has coefficients 1, -1 or 0, it is enough to check that $F_p(X^p)$ is a generator of the module spanned by $\{X^j F_p(X)\}_{j\in \underline{N}}$. Indeed, we have

$$F_{pq}(X)\ F_p(X) = F_p(X^q) \quad . \qquad (10)$$

If $F_p(X^q)$ were not a generator, by theorem 4 there should exist a $f(X^q)$ with coefficients 1 and 0 and such that

$$f(1) < F_p(1) = p \quad .$$

But for a polynomial $g(X)$ belonging to the ideal generated by $F_p(X)$, one has $g(1) \equiv 0 \bmod p$, thus $f(X) = 0$ which proves that $F_p(X^q)$ is a generator.

4. A Class of Unimodular Ideals

We have seen in §3 that for any polynomial $u(X) = \sum_{j<k} X^j$, the ideal $u(X)\underline{Z}[X]$ considered as a module referred to the basis $\{X^j\}_{j\in \underline{N}}$ is unimodular. We shall here recall that some ideals of the cyclic algebra $\underline{Z}[X]$ modulo $X^n - 1$ denoted by A_Z are also unimodular — considered as submodules of $\underline{Z}^n$ referred to its canonical basis — the proof will be found in [1]. If $s \mid n$, the polynomial $X^n - 1/X^{n/s} - 1$ will be denoted by $G_{n,s}(X)$. If I denotes an ideal of A_Z, $\tilde{I}$ will denote the same set considered as a $\underline{Z}$-module.

We have the following theorem.

Theorem 1. *Let t and m be two divisors of n, $f(X) = \text{g.c.d}(G_{n,m}(X), G_{n,t}(X))$. $M = f(X)A_Z$ and $M' = g(X)A_Z$, where $g(X) = X^n - 1/f(X)$. M is the sum of $G_{n,m}(X)A_Z$ and $G_{n,t}(X)A_Z$ and the $\underline{Z}$-modules $\tilde{M}$ and $\tilde{M}'$ are unimodular and orthogonal.*

With the same notations, we obtain the following corollary.

Corollary 1. *Let u be any positive integer. Let $h(X)f(X^u)$ have degree less than nu and non-negative integral coefficients. Then there exist polynomials $k(X)$ and $\ell(X)$ with non-negative integral coefficients such that*

$$h(X)f(X^u) = k(X)\,G_{n,t}(X^u) + \ell(X)G_{n,m}(X^u) \quad .$$

We shall now verify that the lemma of N. G. De Bruijn-Sands [8] is a consequence of corollary 1. This lemma is stated as follows.

If $N = p^\lambda q^\mu M$, where $p^\lambda = n$, $q^\mu = m$ and p and q are primes not dividing M, $A(X)$ is a polynomial of degree less than N with non-negative integral coefficients and $F_{nmd}(X)$ divides $A(X)$ for each divisor d of M, then $A(X)$ can be expressed as

$$A(X) = \frac{X^N - 1}{X^{N/p} - 1} A_p(X) + \frac{X^N - 1}{X^{N/q} - 1} A_q(X)$$

where $A_p(X)$ and $A_q(X)$ are polynomials with non-negative integral coefficients.

We have

$$f(X) = \prod_{d|M} F_{pqd}(X) = \text{g.c.d}(G_{pqM,p}(X), G_{pqM,q}(X))$$

and with $u = p^{\lambda-1} q^{\mu-1}$, $f(X^u) = \prod_{d|M} F_{pqd}(X^u) = \prod_{d|M} F_{nmd}(X)$ and in particular $G_{pqM,p}(X) = \dfrac{X^{pqM} - 1}{X^{qM} - 1}$

$$G_{pqM,p}(X^u) = X^N - 1/(X^{N/p} - 1) \ .$$

<u>Corollary 2.</u> <u>If</u> t <u>and</u> m <u>divide</u> n, <u>the</u> g.c.d <u>of</u> $G_{n,m}(X)$ <u>and</u> $G_{n,t}(X)$ <u>has coefficients</u> 1, -1 <u>and</u> 0 .

This follows from the fact that the polynomial $f(X)$ is a generator of the module M .

<u>Corollary 3.</u> <u>If</u> $n = mt$, $(m,t) = 1$, $f(X) = (X^n - 1)/(X^m - 1)(X^t - 1)$ <u>and</u> $g(X) = (X^m - 1)(X^t - 1)/(X - 1)$, <u>then</u> $f(X)A_z$ <u>and</u> $g(X)A_z$ <u>are unimodular</u>.

Corollary 3 generalizes a result of H. Mann [6], which can be obtained by making m a prime and t a prime.

REFERENCES

[1] P. Camion, "Modules unimodulaires", Journal of Combinatorial Theory 3, to appear.

[2] P. Camion, "Caractérisation des matrices unimodulaires", Cahiers du Centre de Recherche Opérationnelle, vol. 5, no. 4, 1963.

[3] I. Heller, "On linear systems with integral valued solutions", Pacific Journal of Mathem., pp. 1351-1364.

[4] N. G. DeBruijn, "On the factorization of cyclic groups", Indag. Math. Kon. Ned. Akad-Wetensdr. Amsterdam, 15(1953), pp. 370-377.

[5] I. Schoenberg, "A note on the cyclotomic polynomial" Mathematika 11(1964), 131-136.

[6] H. Mann, "On linear relations between roots of unity", Mathematika 12(1965), 107-117.

[7] P. Camion, "Characterization of totally unimodular matrices", Proceedings of the American Mathematical Society, Vol. 16, no. 5, October, 1965.

[8] A. Sands, "On the factorisation of finite abelian groups" Acta Math. Acad. Sci. Hung., 8(1957), 65-86.

Centre d'Informatique
Universite de Toulouse

SOLOMON W. GOLOMB
LLOYD R. WELCH

Algebraic Coding and the Lee Metric

0. Introduction

There is an extensive literature on algebraic coding for the Hamming metric, and relatively little on the corresponding problem for other error metrics.

In the Lee metric for codewords of length n over the alphabet consisting of the integers modulo q, the distance between $(a_1, a_2, \ldots, a_n)$ and $(b_1, b_2, \ldots, b_n)$ is defined to be $\sum_{i=1}^{n} \|a_i - b_i\|$ where $\|a_i - b_i\|$ is the smaller of $a_i - b_i \pmod q$ and $b_i - a_i \pmod q$

In this paper, we consider the existence of close-packed codes in the Lee metric, for general values of q, n, and the "error radius" r. We find that close-packed algebraic codes exist with appropriate values of q: for $n = 1$ and all r; for $n = 2$ and all r; and for $r = 1$ and all n. We conjecture that these are the only cases where close-packed codes exist, and we prove the non-existence of close-packed codes in a variety of cases.

A similar problem, formulated in geometric and algebraic terms, has been considered by S.K. Stein [5]. His geometric shapes also may be regarded as "error spheres", but only in the case of radius 1 are they spheres in the Lee metric. In that case, his construction is identical to ours. More generally, one could define "spheres of radius r in the Stein metric" around the point $A = (a_1, a_2, \ldots, a_n)$, which consist of the intersection of the sphere of radius r in the Lee metric with the sphere of radius 1 in the Hamming metric.

1. The Geometry of Shannon's Five-Phase Code

In [1], Shannon considered the problem of coding to completely eliminate errors in a channel using a 5-symbol alphabet, with the error pattern as shown in Figure 1. The alphabet may be regarded as the integers modulo 5. When the integer r is sent, either r or $r+1$ is received, with respective probabilitites p and q. If one forms a "code" consisting of sending each symbol m times to represent the fact that it occurred once in the message, then there is still a probability of q^m that an error will occur. However, there exists a code using only two code symbols per message symbol which eliminates errors entirely (see Figure 2). In this code, if (a, b) is a codeword, then it may be received as either (a, b) or (a + 1, b) or (a, b + 1) or (a + 1, b + 1). However, we can associate all four of these received messages uniquely with (a, b) when we use the code of Figure 2. This is most readily seen via the geometric representation in Figure 3. The 25 possible codewords (ab) are represented by the 25 cells, with coordinates (a, b). The codewords of Figure 2 correspond to the cells with dots in them. Each dot is in the lower lefthand corner of its "ambiguity square". (The entire 5 × 5 array is to be regarded as a torus.) Since these ambiguity squares are non-overlapping, any received message can be uniquely interpreted.

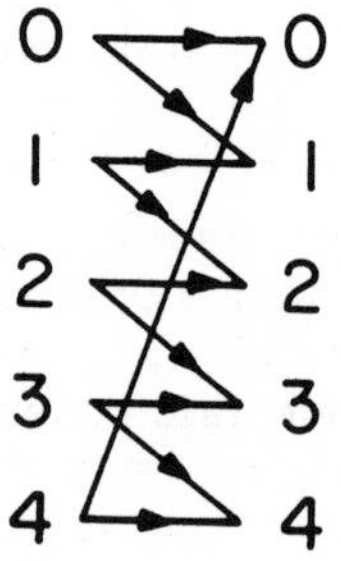

0 = (00)
1 (11)
2 (22)
3 (33)
4 (44)

Figure 1. Shannon's 5-phase channel

Figure 2. An error eliminating code for the channel in Figure 1

The packing of five 2 × 2 squares in the 5 × 5 torus shown in Figure 3 is reasonably efficient, but the resulting code is not close-packed. In particular, there are 5 unused cells in Figure 3. For the channel described by the error statistics of Figure 1, no further improvement is possible. However, if other errors are remotely possible, then it is advantageous to assign the 5 unused squares to the ambiguity regions of the 5 codewords. This can be done as in Figure 4, where the error which occurs when (a, b) is received as (a-1, b) will also be corrected. Since there are no open spaces in Figure 4, this code is "close-packed", and corresponds geometrically to a tiling of the 5 × 5 torus with P-pentominoes.

In general, any tiling of an n × n torus by translations of a given polyomino shape corresponds to a close-packed code, using word-length 2 over the n symbol alphabet. However, the error patterns corrected by such a code are likely to be unnatural or infrequent ones, unless the shape of the polyomino is constrained in various ways. We will next consider a class of polyominoes which satisfy the appropriate constraints.

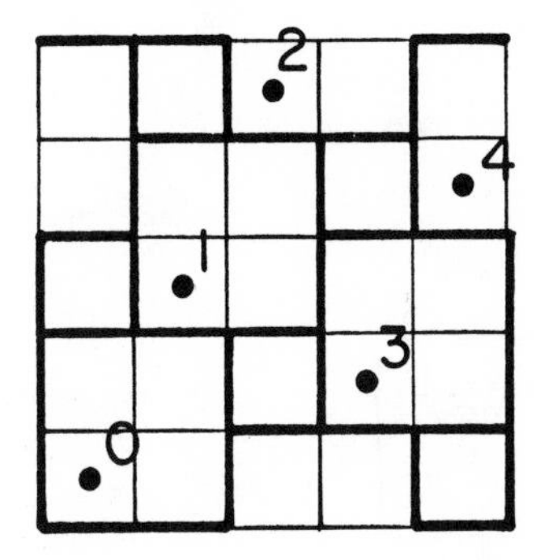

Figure 3. Geometric representation of the code in Figure 2

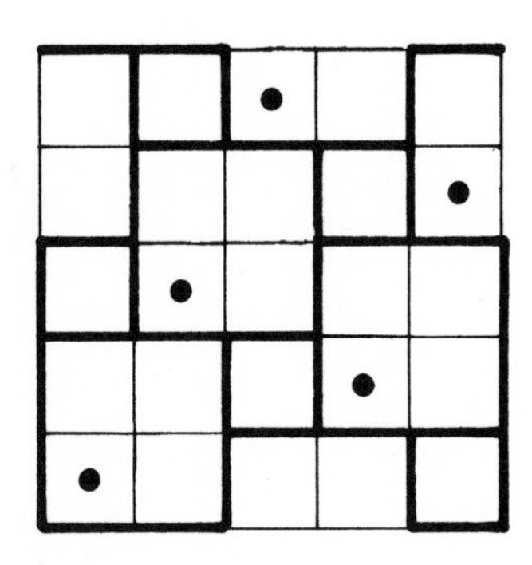

Figure 4. A close-packed P-pentomino code

2. Two-Dimensional Codes in the Lee Metric

In Figure 5, we see the polyomino generated by taking the codeword (a, b) and displacing either component by 1 unit, either up or down. The resulting figure, an X-pentomino, is accordingly a "sphere of radius one" with center at (a, b), in the metric (called the Lee Metric) which computes the sum of the least absolute differences of the corresponding coordinates of two points. (For our purposes, the underlying alphabet is the integers modulo m, and the "least absolute difference" between i and j in this alphabet is the smaller of $i - j \pmod m$ and $j - i \pmod m$.)

In general, a Lee sphere of radius r, in two dimensions, consists of $q = r^2 + (r+1)^2 = 2r^2 + 2r + 1$ cells. The first few cases appear in Figure 6. The obvious close-packed codes to look for are the following ones:

Is there a close-packed code of Lee radius r, with word-length 2, over the q-symbol alphabet, where $q = 2r^2 + 2r+1$?

Such a code would correspond to a tiling of the $q \times q$ torus with polyominoes which are Lee-spheres of radius r. The main result of this section is that such codes exist for all positive integers r.

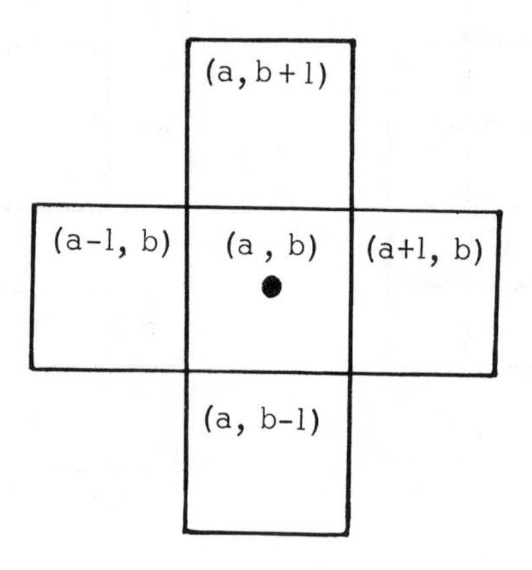

Figure 5. The X-pentomino as a Lee-sphere of radius 1

Theorem 1. For every positive integer r, there is a close-packed r-error-correcting dictionary in the Lee Metric, of codewords of length 2, over the q-symbol alphabet, $q = 2r^2 + 2r + 1$.

Note. Geometrically, this theorem asserts that q Lee-spheres of radius r, in two dimensions, can be used to tile the $q \times q$ torus.

Proof. As codewords, we use the set $\{(q, (2r+1)a)\}$ with $a = 0, 1, 2, \ldots, q-1$, regarding all integers as modulo q. Since these codewords form a group under component-wise addition modulo q, the

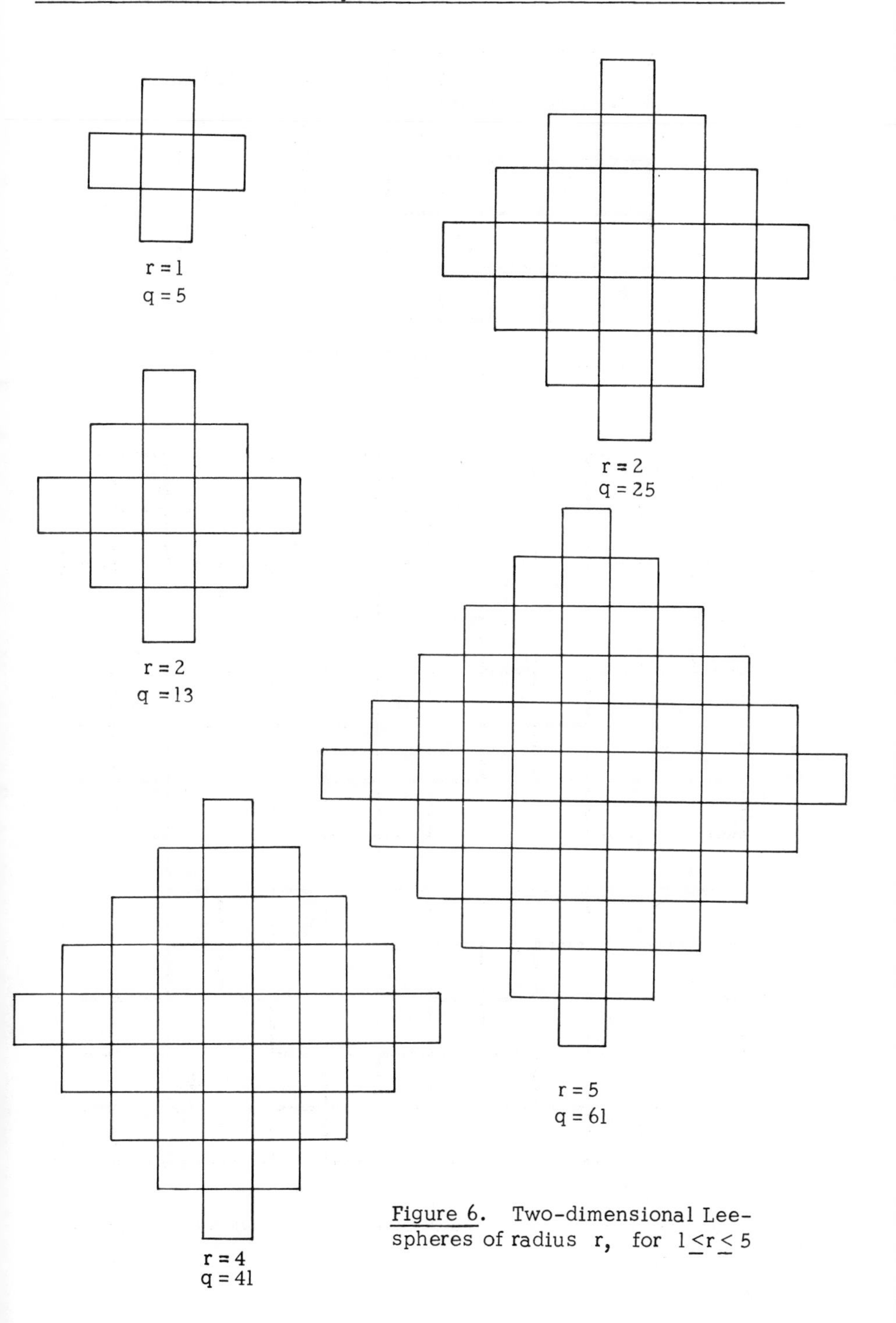

Figure 6. Two-dimensional Lee-spheres of radius r, for $1 \leq r \leq 5$

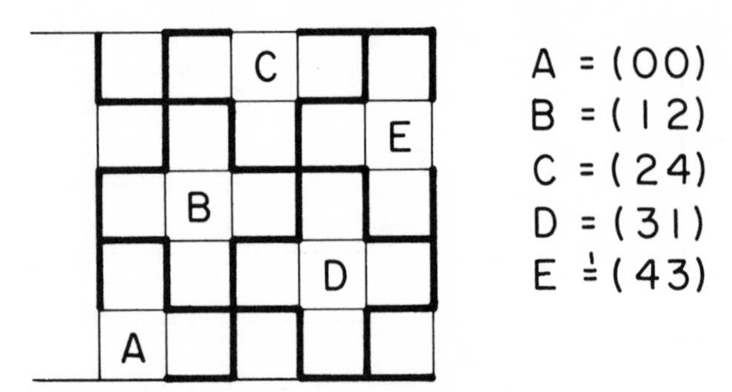

Figure 7. A close-packed code for the Lee metric, using X-pentominoes

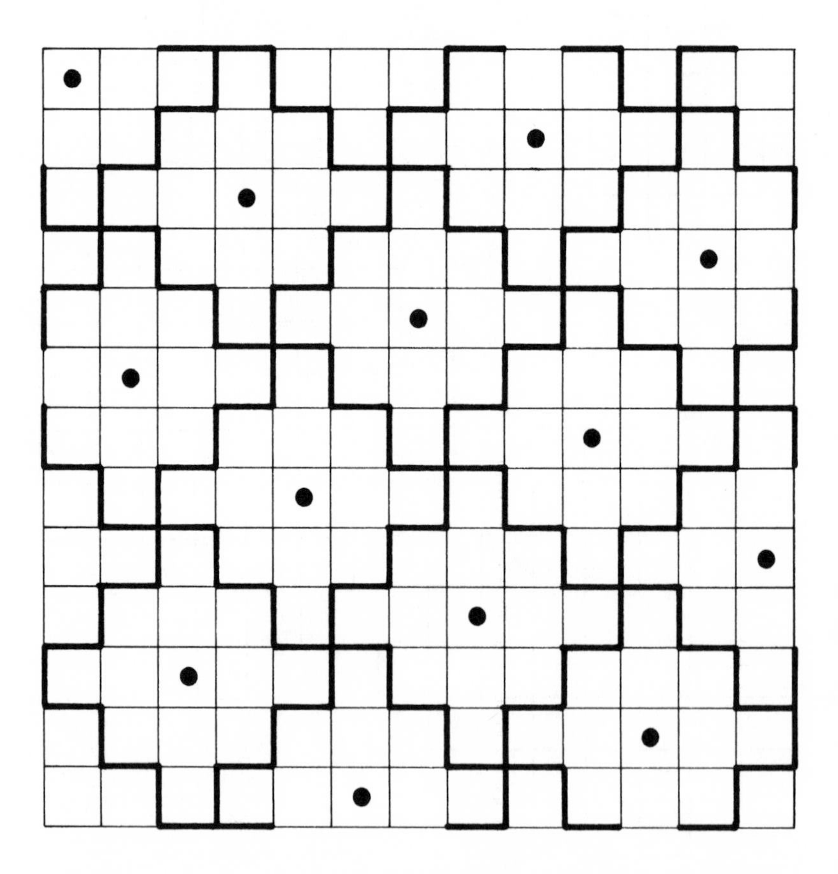

Figure 8. A close-packed double-Lee-error-correcting codes

minimum distance between two codewords equals the minimum weight of any non-zero codeword, and the code is r-error-correcting if this minimum weight is at least $2r + 1$.

Consider $a \not\equiv 0 \pmod q$. If $\|a\| + \|(2r+1)a\| < 2r+1$, then at least one of the two components contributes $\leq r$. If $a = r$, then $(2r+1)a = 2r^2 + r \equiv -(r+1) \pmod q$, so that $\|a\| + \|(2r+1)a\|$ $= r + (r+1) = 2r+1$. More generally, if $a = r - i$ with $0 \leq i \leq r-1$, we have $\|a\| + \|(2r+1)a\| = (r-i) + [(2i+1)r + (i+1)] = 2(i+1)r + 1 \geq$ $2r+1$, as required.

It is not necessary to consider separately the case that the second component is $\leq r$, since $(a, (2r+1)a)$ can be rewritten $(-(2r+1)b, b)$ using $-(2r+1)b = -(2r+1)^2 a = -(4r^2 + 4r + 1)a \equiv a \pmod q$.

Q. E. D.

For $r = 1$, the close-packing of the 5×5 torus with five X-pentominoes (spheres of Lee-radius 1) is shown in Figure 7. Note that the codewords (the centers of the X's) are at the same positions as the codewords in Figures 3 and 4. For $r = 2$, the close-packing of the 13×13 torus with thirteen triskaidekominoes (spheres of radius 2) is shown in Figure 8.

3. Single-Error-Correcting Codes in n Dimensions

A point in n-space has $2n$ other points within a Lee-distance of 1 of it. Geometrically, we may visualize a Lee-sphere of radius 1 in n dimensions as a central hypercube, which has $2n$ hyperfaces, to which another hypercube has been affixed to each of its hyperfaces. The X-pentomino (Figure 5) is the two-dimensional sphere of radius 1. The three-dimensional sphere of radius 1 is the heptacube shown in Figure 9. We can prove directly:

Theorem 2. 49 of the heptacubes of Figure 9 can be used to close-pack the $7 \times 7 \times 7$ hypertorus.

Proof. Specifically, we look at a typical 7×7 cross-section of the solution, shown in Figure 10. The cross-sections of our heptacube will be either X-pentominoes or single squares. In the cross-section shown in Figure 10, we see seven X-pentominoes, as well as seven squares labelled "A" and seven squares labelled "B". The A's are bottoms of heptacubes whose centers are in the plane above, and the B's are tops of heptacubes protruding upward from the plane below. Since the seven A's are systematically translated (1 unit to the northwest) from the centers of the X-pentominoes, we are assured that in the next cross-section above the one we are examining, the X-pentomino sections fit together properly. Similarly the seven B's are systematically translated (1 unit to the southeast) from the centers of the seven X-pentominoes, and are therefore consistent as tops of heptacubes from the layer below. Finally, since 7 is a prime, it is easy to see that these translations must lead to a periodicity of 7 in the third dimension.

Q. E. D.

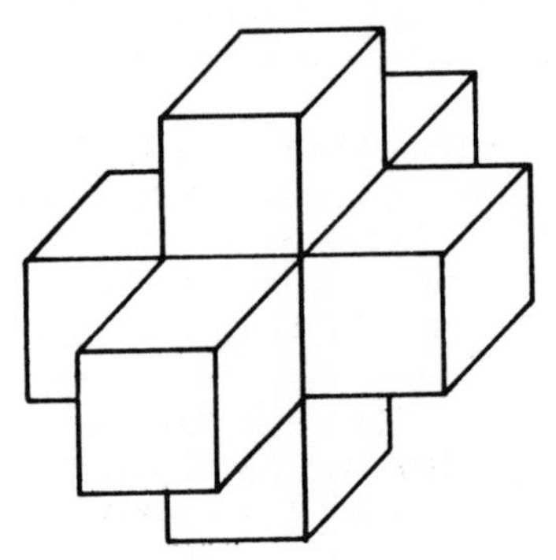

Figure 9. The 3-dimensional Lee-sphere of radius 1

Figure 10. A cross-section of the close-packed $7 \times 7 \times 7$ hyper-torus

A much more general result is true. Basically, it asserts that close-packed single-error correcting codes for the Lee metric exist in n dimensions, for all n, as follows:

Theorem 3. In n dimensions, the spheres of Lee-radius 1 can be used to close-pack the hyper-torus which is $q \times q \ldots \times q = q^n$, where $q = 2n + 1$.

Proof. As centers of the spheres, we use the set S of all points $(a_1, a_2, \ldots, a_n)$ of the hypertorus which satisfy

$$\sum_{i=1}^{n} i a_i \equiv 0 \pmod{2n+1} .$$

The number of solutions to this congruence is clearly q^{n-1}, since an choice of $a_2, a_3, \ldots, a_n$ may be made, and then there is a unique value of a_1, modulo q, to satisfy the congruence. Also, every poin of the hypertorus is within a Lee distance of 1 from some point in thi set. For if $B = (b_1, b_2, \ldots, b_n)$ is any point of the hypertorus, we compute

$$\sum_{i=1}^{n} i b_i \equiv k \pmod{2n+1} ,$$

where $-n \leq k \leq +n$. If $k = 0$, then B is a member of the set S. If $k > 0$, we change b_k to $b_k - 1$ to go from B to a member of S at Lee distance 1 away. If $k < 0$, we change $b_{|k|}$ to $b_{|k|} + 1$ to go from B to a member of S at Lee distance 1 away.

Each point S has only 2n neighbors at a distance of 1 away. Thus the spheres around these points can account for at most $q^{n-1}(2n+1) = q^n$ points if the spheres are all disjoint. However, since every point of the hypertorus is within distance 1 from some point of S, the spheres must be disjoint and fill up the space. Thus, the code is close-packed.

Q.E.D.

4. Some Special Constructions

When $q = 2n + 1$ is a perfect power, it may be possible to construct a close-packed single-error-correcting code in the Lee metric, in n dimensions, with an alphabet size less than q. For example, when $n = 4$ and $q = 9$, rather than tiling the $9 \times 9 \times 9 \times 9$ hypertorus with the spheres of radius 1 composed of 9 tesseracts, as guaranteed by Theorem 3, we may attempt to tile the $3 \times 3 \times 3 \times 3$ hypertorus with such spheres.

A successful attempt at close-packing 9 of these spheres into the $3 \times 3 \times 3 \times 3$ hypertorus is shown in Figure 11. The centers of the spheres are indicated by the boldface letters A through I. The other points of the sphere are indicated by the same letter as the center, but in fainter type. The four coordinates of a point are its 1) superrow, 2) supercolumn, 3) subrow, and 4) subcolumn.

Over the ternary alphabet, a single error in the Lee metric is the same as a single error in the Hamming metric. (If one component is in error, this is a single Hamming error, regardless of the magnitude of the error. For the cyclic ternary alphabet, an error in a component is necessarily by ± 1 modulo 3.) Thus, Figure 11 is also a close-packed single-Hamming-error-correcting code! In fact, this code was obtained in [2] by the method of orthogonal Latin squares. Two orthogonal Latin squares of order n always lead to a single-Hamming-error-correcting code for word length 4 over the n-symbol alphabet as follows:

We label the rows of the squares from 0 to $n-1$, the columns from 0 to $n-1$, and the entries are named 0 to $n-1$. Then we form the set of all quadruples

$$(r, c, e_1, e_2)$$

where r is the row index, c is the column index, e_1 is the entry at the (r, c) position in the first square, and e_2 is the entry at the (r, c) position in the second squares. It is easy to show that if (r, c, e_1, e_2) and (r', c', e_1', e_2') agree in any two of their components, then they must agree in all four. Hence, the set of all n^2 "points" (r, c, e_1, e_2) has a minimum Hamming distance of 3 between any two members, and is therefore single-error-correcting.

The case n = 3 is illustrated in Figure 12. The code obtained is the same as in Figure 11.

In the Hamming metric, a "sphere of radius r" looks even less "round" than a sphere in the Lee metric. In [2], these Hamming-metric "spheres" are designated as rook domains. Specifically, in two dimensions, the single Hamming errors from the point (a, b) correspond to those squares to which a rook, located on the square (a, b) could go in a single move. (See Figure 13.) In 2 dimensions, rook domains do not pack efficiently, but in higher dimensions they may. Besides the theory of rook-domain packing in [2], there is also a fundamental outstanding conjecture [3].

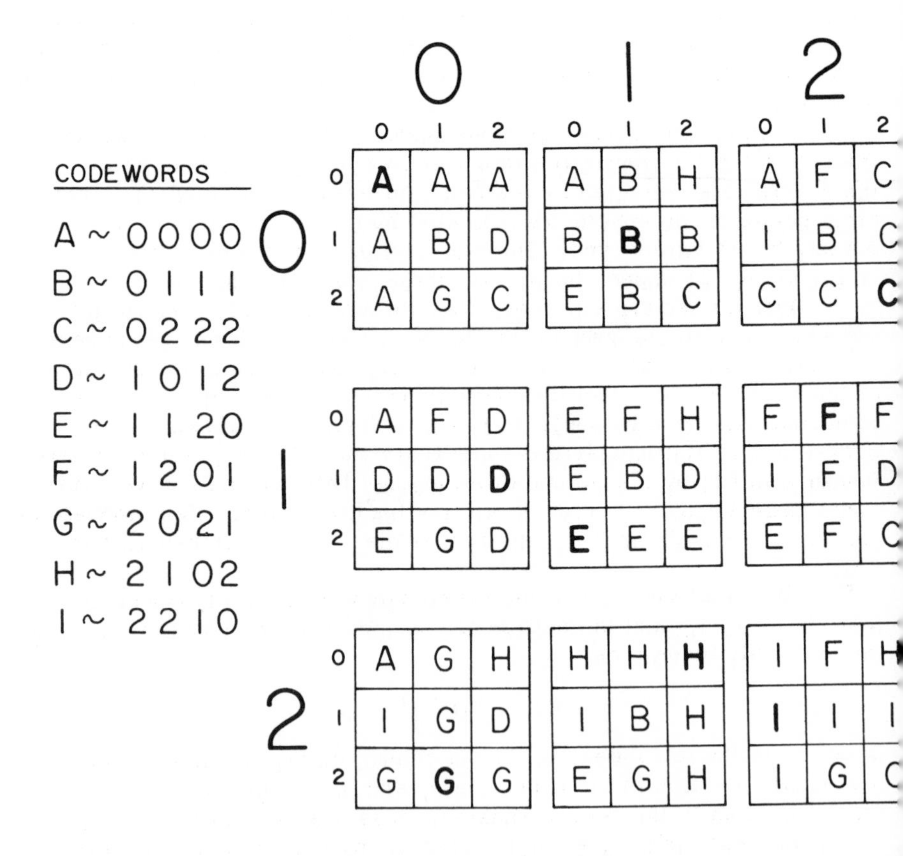

Figure 11. Close-packed four-dimensional code, single-error correct in both Hamming and Lee Metrics

r	c	e_1	e_2
0	0	0	0
0	1	1	1
0	2	2	2
1	0	1	2
1	1	2	0
1	2	0	1
2	0	2	1
2	1	0	2
2	2	1	0

	0	1	2
0	0	1	2
1	1	2	0
2	2	0	1

0	1	2
0	1	2
2	0	1
1	2	0

Figure 12. From orthogonal Latin squares to a distance 3 code

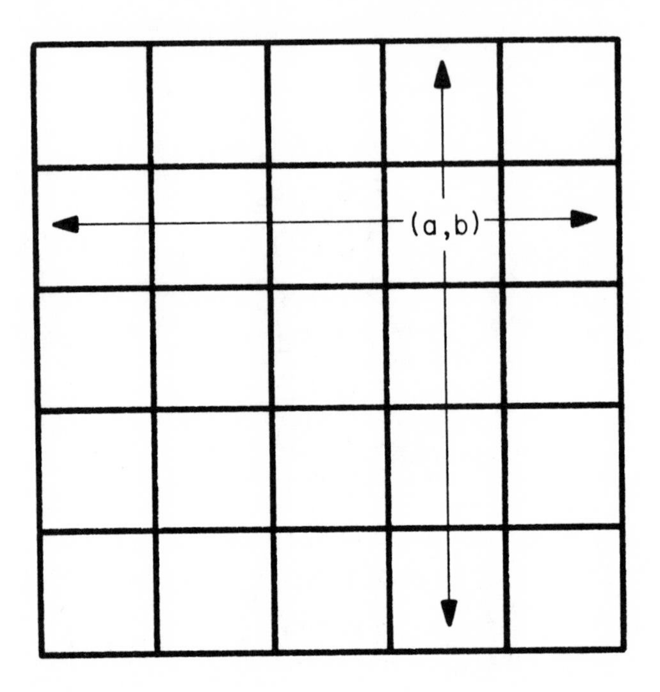

Figure 13. The rook domain of the square (a, b)

5. Sphere-Packing Constraints

If we denote by $V(n,r)$ the number of points contained in the n-dimensional sphere of Lee-radius r, it is rather easily established (see the proof of Theorem 4, below) that

$$V(n,r) = \sum_{k\geq 0} 2^k \binom{n}{k}\binom{r}{k} .$$

It is a curious fact that $V(n,r)$ is symmetric in n and r. The effective upper limit of the summation is at $k = \min(n,r)$. By the usua sphere-packing argument, we obtain the following "sphere-packing bound":

Theorem 4. The number of codewords in an r-error-correcting code dictionary, for word-length n and alphabet size q, $(q \geq 2n+1)$, where errors are measured in the Lee metric, cannot exceed

$$q^n / \sum_{k\geq 0} 2^k \binom{n}{k}\binom{r}{k} .$$

Proof. The codewords must be surrounded by disjoint spheres of radius r. There are q^n points in the sphere, and each codeword uses up $V(n,r)$ of them, so that there can be at most $q^n/V(n,r)$ codewords.

To establish the identity

$$V(n,r) = \sum_{k \geq 0} 2^k \binom{n}{k} \binom{r}{k}$$

we regard the n components of a codeword as "boxes", and we have r "error balls" to distribute among these boxes. For any $k \leq r$, we consider the problem of distributing up to r error balls into exactly k boxes. There are $\binom{n}{k}$ ways to choose k of the n boxes to contain all the balls; each of these k boxes must be designated as either containing a positive or negative deviation, for a factor of 2^k; and there are $\binom{r}{k}$ ways to distribute up to r balls into k boxes in such a way that no box is empty. Multiplying these three factors together, and then summing over k, leads directly to the formula for $V(n,r)$. Q.E.D.

A close-packed code is one which attains the sphere-packing bound. Clearly, a necessary condition for a close-packed code to exist, for given n, r, and q, is that $V(n,r)$ divide q^n. This necessary condition is met, in particular, when $q = V(n,r)$. However, as we shall see, $q = V(n,r)$ is both unnecessary and insufficient for a close-packed code to exist.

The underlying geometric problem is this: For what values of n and r does the n-dimensional sphere of radius r, $S_{n,r}$, tile n-dimensional space? If the sphere is incapable of tiling the space, then no close-packed code can exist. If the sphere does tile the space, then any q such that the tiling is periodic with period q in each direction is an acceptable alphabet size.

The spheres $S_{1,r}$ and $S_{2,r}$ all succeed in filling their respective spaces, since $S_{1,r}$ is simply a line segment of length $2r+1$, which fills up one-dimensional space with a periodically $q = 2r+1$; and $S_{2,r}$ is the two-dimensional sphere of Theorem 1, which fills the plane periodically with a period of $q = 2r^2 + 2r + 1$.

Also, the spheres $S_{n,1}$ fill up n-dimensional space, according to Theorem 3, with a periodicity of $q = 2n + 1$. A smaller q (specifically, a factor of $2n + 1$ containing all the distinct prime factors of $2n + 1$) may sometimes be possible, as illustrated for $n = 4$, $q = 3$ in Figure 11.

Not all spheres $S_{n,r}$ are capable of tiling n-dimensional space. The first counter-example is:

Theorem 5. The sphere $S_{3,2}$, illustrated in Figure 14 and made up of 25 unit cubes, is unable to tile 3-space.

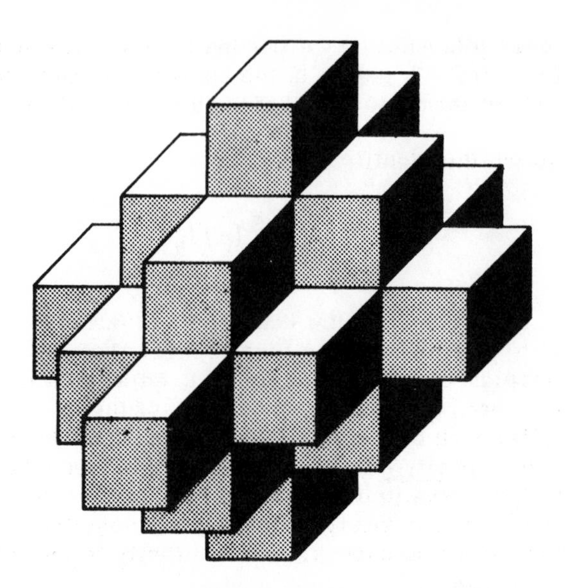

Figure 14. The Lee-sphere $S_{3,2}$, composed of 25 unit cubes

Proof. Let $S(a,b,c)$ be the Lee sphere of radius 2, dimension 3 and center (a,b,c). Assume that E^3 can be tiled and let

$$\{S(a_i, b_i, c_i) \mid i = 0,1,2,\ldots\}$$

be a tiling. We may also assume $(a_0, b_0, c_0) = (0,0,0)$.

Let $S(a_1, b_1, c_1)$ be the sphere containing $(2,1,0)$, so that

$$|a_1 - 2| + |b_1 - 1| + |c_1| \leq 2 .$$

Since $S(a_1, b_1, c_1)$ and $S(0,0,0)$ are disjoint,

$$5 \leq |a_1| + |b_1| + |c_1| .$$

However, by the triangular inequality,

$$|a_1| + |b_1| + |c_1| \leq |a_1 - 2| + 2 + |b_1 - 1| + 1 + |c_1| \leq 2 + 3 = 5 ,$$

with equality holding if and only if $a_1 \geq 2$, $b_1 \geq 1$. It follows that $a_1 \geq 2$ and $b_1 \geq 1$ and $a_1 + b_1 + |c_1| = 5$.

If $a_1 \geq 3$ then

$$|a_1 - 3| + |b_1| + |c_1| = a_1 - 3 + b_1 + |c_1| = 2$$

and $(3,0,0) \in S(a_1, b_1, c_1)$.

The point $(2, -1, 0)$ is outside S_0 and S_1 and therefore in a third sphere S_2 . An argument similar to the one above shows that $a_2 > 2$ and if $a_2 \geq 3$ then $(3,0,0) \in S_2$. Since S_1 and S_2 are disjoint, either $a_1 = 2$ or $a_2 = 2$. Using a symmetry of E^3, we may assume $a_1 = 2$ and $c_1 \geq 0$, and consider the three cases:

1) $(a_1, b_1, c_1) = (2,3,0)$

2) $(a_1, b_1, c_1) = (2,2,1)$

3) $(a_1, b_1, c_1) = (2,1,2)$

Again a symmetry of E^3 can be used to reduce case 3) to case 2) .
Case 1).

$S_0 = S(0,0,0)$ and $S_1 = S(2,3,0)$ are members of a tiling. Since the point $(1,1,1)$ is not in S_0 or S_1, it must be in another S, say $S(a_2, b_2, c_2)$. We have:

IE (1)

$$|a_2| + |b_2| + |c_2| \geq 5$$

$$|a_2 - 2| + |b_2 - 3| + |c_2| \geq 5$$

$$|a_2 - 1| + |b_2 - 1| + |c_2 - 1| \leq 2 \ .$$

The only solution to these inequalities is

$$(a_2, b_2, c_2) = (1,1,3) \ .$$

Next, consider the point $(1,2,1)$ and the linear transformation of E^3 $\varphi(x,y,z) = (2-x, 3-y, z)$. This maps $(1,2,1)$ into $(1,1,1)$ and interchanges S_0 and S_1 . Therefore the center of the sphere, S_3, containing $(1,2,1)$ is $\varphi^{-1}(1,1,3) = (1,2,3)$. S_2 and S_3 are neither disjoint nor identical, contrary to the initial hypothesis.
Case 2).

$S_0 = S(0,0,0)$ and $S_1 = S(2,2,1)$ are members of a tiling.

Consider the point $(1,1,-1)$. Using an argument similar to Case 1 we have

IE 2)

$$|a_2| + |b_2| + |c_2| \geq 5$$

$$|a_2 - 2| + |b_2 - 2| + |c_2 - 1| \geq 5$$

$$|a_2 - 1| + |b_2 - 1| + |c_2 + 1| \leq 2$$

The only solution to these inequalities is

$$(a_2, b_2, c_2) = (1, 1, -3) \ .$$

The point $(1, 2, -1)$ is not in S_0, S_1 or S_2 and therefore in some S_3. We obtain the inequalities

IE 3)
$$|a_3| + |b_3| + |c_3| \geq 5$$
$$|a_3 - 2| + |b_3 - 2| + |c_3 - 1| \geq 5$$
$$|a_3 - 1| + |b_3 - 1| + |c_3 + 3| \geq 5$$
$$|a_3 - 1| + |b_3 - 2| + |c_3 + 1| \leq 2$$

The second and fourth inequalities imply

$$a_3 \leq 1, \quad c_3 \leq -1$$

while the third and fourth imply

$$b_3 \geq 2, \ c_3 \geq -1, \ \text{and} \ |a_3| + b_3 - 2 + |c_3 + 3| = 5 \ .$$

With this information, the first and fourth then imply the unique solution

$$(a_3, b_3, c_3) = (1, 4, -1) \ .$$

Next, let φ be the mapping of E^3 where $\varphi(x, y, z) = (z + 3, y - 1, x - 1)$. The tiling produced by applying φ to the hypothesized tiling has, as members, $\varphi S_2 = S(0, 0, 0)$ and $\varphi S_3 = S(2, 3, 0)$. But this is Case 1) which has already been shown to yield a contradiction.

Q. E. D.

6. The Equivalent Tesselation With Cross-Polytopes

A general proof of the inability of $S_{n,r}$ to tile n-dimensional space, for large classes of n and r, can be based on the approximation of $S_{n,r}$ by the n-dimensional cross-polytope.

Def. For every Lee-sphere $S_{n,r}$, we define the conscribed cross-polytope to be the smallest convex figure containing the 2^n center points of its (n-1)-dimensional extremal hyper-faces.

In Figure 15, the conscribed cross-polytopes are illustrated in 2 and 3 dimensions. In 2 dimensions the figure is a square, and in 3 dimensions, a regular octahedron. In n dimensions, it is the regular cross-polytope, of n-dimensional hyper-volume

$$V_{CP}(n, r) = \frac{d^n}{n!} = \frac{(2r+1)^n}{n!} \ ,$$

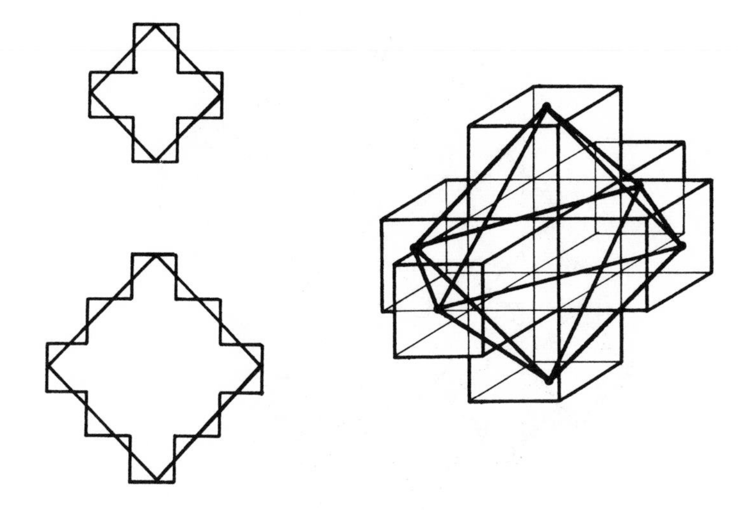

Figure 15. Examples of the conscribed cross-polytopes in 2 and 3 dimensions

where $d = 2r + 1$ is the Euclidean diameter of $S_{n,r}$.

The significant fact about these figures is that any tiling of n-dimensional space with the spheres $S_{n,r}$ induces a (less efficient) tiling with the conscribed cross-polytopes. In general, the relative efficiency factor is

$$\frac{V_{CP}(n,r)}{V(n,r)}$$

which is less than unity whenever $n > 1$.

In Figure 16 we see a tiling of the plane with the X-pentomino ($S_{2,1}$) and the induced tessalation with conscribed squares. The efficiency of this square tiling is:

$$\frac{V_{CP}(2,1)}{V(2,1)} = \frac{9/2}{5} = 90\% ,$$

and we observe that for each square of area $9/2$ conscribed in a pentomino, there is a left-over square of area $1/2$.

We use this type of argument to prove the following two theorems.

Theorem 6. The sphere $S_{3,r}$ cannot tile 3-space for any $r > r_0$.

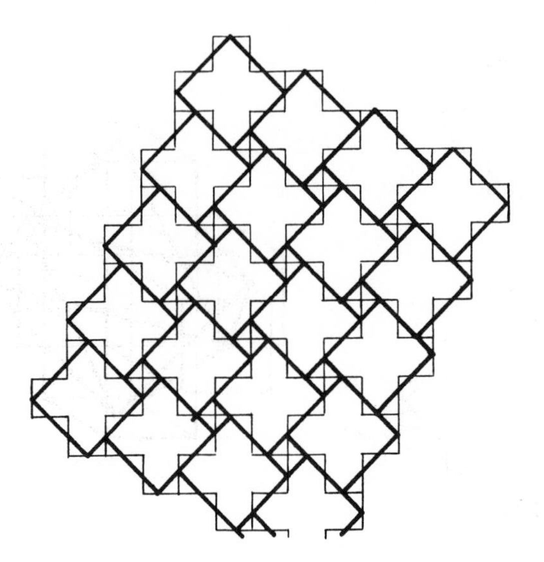

Figure 16. The conscribed square tiling induced by the $S_{2,1}$ tiling

Proof. If $S_{3,r}$ tiles 3-space, it induces a packing of 3-space with (conscribed) regular octahedra, with a packing efficiency of

$$E_3(r) = \frac{V_{CP}(3,r)}{V(3,r)} = \frac{(2r+1)^3/6}{(8r^3+12r^2+16r+6)/6} = \frac{(2r+1)^3}{(2r+1)^3+5(2r+1)} = 1/(1+5/(2r+1)$$

Now it is known [4] that equal regular octahedra are not capable of completely filling three-dimensional space. It can be shown that if a figure does not fill space with a packing efficiency of unity, then there is an upper bound α to the packing density, with $\alpha < 1$. (For the octahedron, there is an obvious construction with a packing efficiency of 3/4, but we have not located a reference listing the best possible α.) As soon as E_3 exceeds α, the Lee-sphere packing induces an octahedral packing which exceeds the limit on octahedral packing density. (If indeed $\alpha = 3/4$, then at $r = 1$, $E_3 = 9/14 < 3/4$, but at $r = 2$, $E_3 = 5/6 > 3/4$, so that no close-packing of 3-space by Lee spheres of radius $r > 1$ would be possible.) Since $E_3(r) \to 1$ as $r \to \infty$, $E_3(r) > \alpha$ for $r > r_0$. Q.E.D.

This is readily generalized as follows:

Theorem 7. For $n > 2$ and $r > \rho_n$, the sphere $S_{n,r}$ cannot tile n-space.

Proof. In n-dimensional Euclidean space, for $n > 2$, it is known [4] that the regular cross-polytope does not tile the space. Again, there is a maximum packing density α_n, which would be exceeded by the conscribed spheres, for $r > \rho_n$, if the Lee-sphere packing existed. This depends only on the fact that

$$E_n(r) = \frac{V_{CP}(n,r)}{V(n,r)} \to 1 \text{ as } r \to \infty .$$

Q.E.D.

Again, references to limiting packing densities for the cross-polytopes have not been found.

7. Summary

The Lee spheres $S_{n,r}$ are found to tile n-dimensional Euclidean space, in close-packed fashion, when:

a) $n = 1$, for all r

b) $n = 2$, for all r

c) $r = 1$, for all n .

It is conjectured that these are the only cases for which a close-packing exists. The close-packing has been proved not to exist when:

a) $n = 3$, $r = 2$

b) $n > 2$, $r > \rho_n$, where ρ_n depends on the limit to packing efficiency of the cross-polytope in n-dimensional Euclidean space.

ACKNOWLEDGEMENTS. We wish to acknowledge some very helpful suggestions made by E. Berlekamp and by R. K. Guy.

REFERENCES

[1] Shannon, C. E., "The Zero Error Capacity of a Noisy Channel", IRE Trans. on Information Theory, Vol. IT-2, No. 3; pp. 8-16, September, 1956.

[2] S. W. Golomb and E. C. Posner, "Rook Domains, Latin Squares, Affine Planes, and Error-Distributing Codes", IEEE Transactions on Information Theory, Vol. IT-10, No. 3, July 1964; pp. 196-208.

[3] S. W. Golomb and E. C. Posner, "Hypercubes of Nonnegative Integers", Bulletin of the American Mathematical Society, Vol. 71, 1965; pg. 587.

[4] H. S. M. Coxeter, Regular Polytopes , The Macmillan Company New York, Second Edition, 1963.

[5] Sherman K. Stein, "Factoring by Subsets," Pacific Journal of Mathematics, vol. 22, no. 3, 1967, pp. 523-541.

R. TURYN

Sequences With Small Correlation

The purpose of this paper is to present certain results and techniques which center around the problem of finding sequences with small correlation functions. The paper is concerned with the exposition of the mathematical problems. The sequences find applications as near-orthogonal codes and in phase modulation in synchronization and pulse compression systems. They may be viewed as codes designed to correct errors in separation of successive words, with a distance criterion which is not the Hamming metric.

The general subject is one of finding complex functions $f(t)$ such that $c(\tau) = \int f(t)\ \bar{f}(t+\tau)\,dt$ is small except near $\tau = 0$. The problems arise in specifying the form of the function $f(t)$, the meaning of the integral sign, and of course in finding the function f. In the following, $f(t)$ will be assumed to be constant on successive equal time intervals, and the values of $f(t)$ will be assumed to be roots of 1, so that $f(t)$ will be represented by a finite sequence of roots of 1.

1. The binary non-periodic case

Let $x_1, \ldots, x_n$ be a sequence of complex numbers, and

$$c_j = \sum_1^{n-j} x_i\, \bar{x}_{i+j}$$

$$a_j = c_j + \bar{c}_{n-j} = \sum_1^{n} x_i\, \bar{x}_{i+j}$$

with $i+j$ reduced $\mod n$ if necessary in the last expression. The c_j are the non-periodic, the a_j the periodic, correlation function. Then we have

$$\begin{aligned} |\textstyle\sum_1^n x_i|^2 &= \textstyle\sum_1^n |x_i|^2 + \sum_1^{n-1} a_i \\ &= c_0 + 2\textstyle\sum_1^{n-1}(c_i + \bar{c}_i) \quad . \end{aligned} \tag{1}$$

The first problem we wish to consider is that of making the c_i small with $x_i = \pm 1$. The following facts are then obvious:

1. c_j is an integer with parity of $n-j$.

2. By equation (1), $n + \sum_1^{n-1} a_i = (\sum x_i)^2 > 0$ implies that the average of the $a_i \geq -1$.

3. By letting $x_i = 1 - 2y_i$, we see that

$$a_i \equiv n \pmod 4 \quad .$$

It follows that the most stringent uniform condition we can impose on the c_i is that $c_i = 0$ or -1 for all i $(1 \leq i \leq n-1)$.

This was the condition imposed by Barker, who found sequences with these properties of lengths 3, 7, 11. It has become customary to refer to Barker sequences as those binary sequences which satisfy $|c_j| \leq 1$, for all j.

It was shown (Storer and Turyn, and Poliak and Moshetov) that there are no other Barker sequences of odd length. The proof of the theorem shows why it is difficult to extend it. The obvious procedure is to take a Fourier transform of the correlation integral, and find solutions of

$$\hat{f}\bar{\hat{f}} = \hat{c}(\tau)$$

($\hat{f}$ = Fourier transform, or some similar transform, of f). In the non-periodic case, f is simply a polynomial in an indeterminate, $\bar{f}$ the polynomial with roots $(\bar{\alpha}^{-1})$ if f has roots (α). However, even if it is possible to find a suitable factorization of $c(\tau)$ given a reasonable function $\hat{c}(\tau)$, it seems impossible in practice to determine whether the resulting polynomial f, (assuming its roots known) will have all coefficients ± 1. This approach is very useful when the integral is over a finite abelian group (e.g., if the a_j are prescribed, the cyclic group), as we shall show later. However, the following method does work to some extent: the parity of the c_j, the fact that $c_j = 0, \pm 1$ and the equation $a_j \equiv n \pmod 4$ immediately determine the a_j completely except when $n \equiv 2 \pmod 4$, (and then it is very easy to see that the condition $|c_j| \leq 1$ implies $n = 2$). Further, if n is odd, $c_j + c_{n-j} = \pm 1 = (-1)^{(n-1)/2}$ (the remainder of n after division by 4) so that $c_j = 0$ for $n-j$ even, $c_j = (-1)^{\frac{n-1}{2}}$ for $n-j$ odd, and the c_j are in fact unambiguously determined by the condition $|c_j| \leq 1$. For $n \equiv 0 \pmod 4$ we get $a_j = 0$, i.e., $c_j = -c_{n-j}$, $c_j = \pm 1$ for j odd, but the sign of c_j is not determined completely. Next, we have the equations

$$\prod_{1}^{n-k} x_i x_{i+k} = (-1)^{(n-k-c_k)/2} \qquad (2)$$

by a simple count in the definition of c_k, which yields

$$x_{k+1} x_{n-k} = (-1)^{n-k-(c_k + c_{k+1} + 1)/2} \tag{3}$$

(the "reflection rule"). The next step in the odd length theorem is to put this information back into the defining equations for the c_k, try the counting trick once more, and to see how the equations work. The point is that the c_j must be completely determined for this procedure. Note, in passing that (3) becomes, for the odd length Barker sequences,

$$x_{k+1} x_{n-k} = (-1)^{1+k+(n+1)/2} \tag{4}$$

so that the products $x_i x_{n+1-i}$ are alternately +1 and -1. It is easy to verify that sequences which satisfy this last condition $(x_k x_{n+1-k} = u(-1)^k)$ also satisfy $c_j = 0$ for $n-j$ even, since the terms in c_j will cancel in pairs (first and last, etc.).

There is overwhelming evidence that there are no Barker sequences with n even, $n > 4$. D. J. Newman (unpublished) has shown that the finiteness of the number of Barker sequences would follow from the conjecture:

$$\int_0^1 \left|\sum_1^n x_m e^{2\pi imx}\right| dx \leq \sqrt{n-1}$$

for large n. (Indeed, letting $F(x) = \sum_m x_m e^{2\pi imx}$, we have

$$\int_0^1 |F|^4 dx = n^2 + 2\sum c_j^2 = n^2 + 2\left[\frac{n}{2}\right] < \frac{n^3}{n-1} .$$

But by the Minkowski inequality

$$\int |F|^2 dx \leq \left(\int |F|^4 dx\right)^{1/3} \left(\int |F| dx\right)^{2/3}$$

so that $\int_0^1 |F|^2 dx = n$ implies

$$n < \left(\frac{n^3}{n-1}\right)^{1/3} \cdot (n-1)^{1/3} = n .$$

Note that this proof uses the fact that half of the c_i are 0, the others ± 1. It is possible that sharper estimates for $\int |F| dx$ and a sharper inequality than Minkowski's might yield better estimates. I do not know of any sharp asymptotic estimates for $b(n) = \underset{n}{\text{Min}} \underset{j>0}{\text{Max}} |c_j|$ (the minimum over all binary sequences of length n); in fact, the existence of only a finite number of Barker sequences ($|c_j| \leq 1$) has not yet been proved. There are results of Moser and Moon (to appear) and I. Jacobs (unpublished) that, roughly, with probability approaching 1, $\underset{j}{\text{Max}} |c_j|$ is of the order of $\sqrt{n}$ if a binary sequence is picked at random (+1 and -1 equally likely, successive choices independent). These results show that the percentage of sequences which satisfy $|c_j| \leq B$ goes down as n increases, but do not show that the number of such sequences decreases. A plausible conjecture is that $b(n)$ approaches infinity with n, perhaps like $\log n$.

An exhaustive computer search has shown that for $n \leq 28$, there are sequences of length ≤ 21, 25 and 28 with $|c_j| \leq 2$ for all j, but none of length 22, 23, 24, 26, 27. The autocorrelation functions of these sequences do not seem to reveal any pattern or suggest any plausible conjectures: the procedure outlined above could be used to search for binary sequences with specified c_j, if a plausible conjecture could be made. An obvious approach is to try to make up long sequences using short ones, but no one yet has found a construction which diminishes $\underline{\text{Max}|c_j|}_{n}$. The most obvious approach is to use the tensor product: if $X = (x_i)$, $Y = (y_j)$ are two sequences, of lengths m and n respectively, we form the product sequence as

$$X \cdot Y = Xy_1, Xy_2, Xy_3 \ldots$$

$$\text{or } X \cdot Y = (Z_j) \text{ with } Z_{rn+s} = x_s y_{r+1} .$$

We then have the formula

$$C_{qn+r}(X \cdot Y) = C_r(X) \; C_q(Y) + \bar{C}_{n-r}(X) \, C_{q+1}(Y) \quad 0 \leq q, \; r < n$$

from which we get

$$a_{qn+r}(X \cdot Y) = C_r(X) \; a_q(Y) + \bar{C}_{n-r}(X) \; a_{q+1}(Y)$$

which shows that the tensor product cannot diminish $\underline{\text{Max}|c_j|}_{n}$ and suggests that if $\max|c_j|$ is approximately $(\text{length})^{1/2}$ for both X and Y, it is also about that for X Y. The tensor product of two sequence is really defined on the product of two groups, and this is a way of shifting the definition to a different group.

An interesting question related to the behavior of $b(n)$ is the behavior of $\min_n \sum c_i^2$ (again, the minimum over all binary sequences of length n). Lunelli has tabulated some of the sequences which attain the minimum for small n. The Barker sequences, when they exist, clearly furnish a minimum for this problem also. Intuitively one would expect this problem to be easier, and the solutions to $\min_n \max_j |c_j|$ to be close to $\min_n \sum c_i^2$, and vice-versa, at least for small n. However, I do not know of any results on this problem (other than the tabulation for small n).

2. The binary periodic case

We now turn our attention to binary sequences defined on the cyclic group Z_v, and the problem of finding sequences with $|a_j|$ small. The known Barker sequences arise from difference sets, with $a = v - 4n = -1$ for $v = 3, 7, 11$, $a = 0$ for $v = 4$ and $a = 1$ for $v = 5, 13$. (Here v denotes the length.) The difference sets with $a = -1$ are known as Hadamard difference sets, since the incidence matrix of the design becomes a Hadamard matrix when bordered suitably. $a = 0$ ($v = 4n$, $n = N^2$,) corresponds to unnormalized Hadamard matrices. It has been conjectured by Ryser that no such cyclic difference sets ($a = 0$, circulant Hadamard matrices) exist for $v > 4$. This would imply, of course, that there are no Barker sequences of length > 13. The problem was extensively considered in Turyn [1], and the conjecture seems eminently plausible (the known facts are reviewed below; there are no such sets for $1 < N < 55$). $v - 4n = 1$ occurs for $v = 5$ (a trivial set, $k = n = 1$) and $v = 13$, which can be viewed as a fourth-power difference set modulo a prime, or as the geometry of the plane over $GF(3)$. $a = 1$ implies that $v = 2m^2 + 2m + 1$, $k = m^2$, $n = \frac{m(m+1)}{2}$, $\lambda = \frac{m(m-1)}{2}$. There are no such sets for $3 \le m \le 11$. $a = 2$ implies that $v - 4n = 2$, so that n must be a square, and $v = 4m^2 + 2$, $n = m^2$, $k = 2m^2 + 1 - \sqrt{3m^2 + 1}$ and thus $3m^2 + 1 = h^2$. m and h are determined recursively by the equations

$$h_{j+1} = 2h_j + 3m_j, \quad m_{j+1} = 2m_j + h_j \quad .$$

$h = 2$, $m = 1$ gives $n = 1$, $v = 6$ which is a trivial case. The next two values are $v = 66$, $n = 16$ and $v = 902$, $n = 225$, and it is easy to verify that there are no difference sets with these parameters. $a = 3$ implies that $2k = 3 + 4n - \sqrt{16n + 9}$, so $v = s^2 + s + 1$, $n = 1/4\,(s+2)(s-1)$, $k = \frac{v-(2s+1)}{2} = \frac{s(s-1)}{2}$. Here again it can be verified that there are no such sets for $s = 5, 6, 9, 10, 13, 14, 17, 18$ (the first 8 values of s which might be possible). $a = 4$ implies $n = N^2$, $v = 4N^2 + 4$, $k = \frac{v}{2} - b$, with $5N^2 + 4 = b^2$. Here $N = 1, b = 3$ is a solution (trivial set, $k = n = 1$) and N_{j+1}, b_{j+1} are determined from N_j, b_j recursively by the equations

$$N_{j+1} = \frac{1}{2}(b_j + N_j) \quad b_{j+1} = N_{j+1} + 2N_j$$

$N = 3$ corresponds to $v = 40$, $n = 9$ ($v = 3^3 + 3^2 + 3 + 1$, the 3 dimensional geometry over $GF(3)$). The next possible value is $N = 8$, $v = 260$ and there is no difference set with these parameters.

Thus, the cyclic difference sets with $v - 4n = a$ small are scarce, with the possible exception of the Hadamard sets, for which $a = -1$. There are four classes of these which are known: the quadratic residues for primes $\equiv -1 \pmod 4$, the Hall sets for primes of the form $4x^2 + 27$ (which are $\equiv -1 \pmod 4$), the Brauer sets of length $p_1 p_2$ with the p_i twin primes ($p_1 - p_2 = 2$) and, finally, the Singer sets for $p = 2$ and the Gordon - Mills - Welch generalization. The last class arises from the theorem that the Kronecker product of the matrix $\begin{pmatrix} 1 & 1 \\ 1 & -1 \end{pmatrix}$ with itself n times is an Hadamard matrix, whose rows and columns can be identified in a natural way with all the points of n-dimensional affine space over $GF(2)$. Singer's construction shows that the core of this matrix can be rearranged to be cyclic. This class of sequences was rediscovered by Golomb and Zierler, who concentrated on the linear recursion. (Singer enumerated the non-zero elements of the vector space by taking the powers of a primitive element α of the extension of degree n of the base field, here $GF(2)$; the maximal length linear shift register sequences arise by taking $\ell_0(\alpha^i)$, where ℓ_0 is any non-trivial linear functional into the base field.) The Gordon - Mills - Welch sequences arise from Singer sequences by a process which can be described as dividing a long one by a corresponding short one, and re-multiplying by any sequence of the short length which also has $a_j = -1$ for all $j \neq 0$. The lengths of the Gordon - Mills - Welch sequences (with $a = -1$) are thus all $2^n - 1$. In addition to these, Baumert found three sequences of length 127 with $a_j = -1$ which are not in any of the above four categories. Not only are the difference sets inequivalent, but the (127, 63, 31) designs they define are non-isomorphic (the counts of the sizes of the triple intersections are different). Baumert has found no other such sets for $k \leq 100$. It is probable that there are some which are not gotten from the Gordon - Mills - Welch construction for $v = 255$, $k = 127$.

D. H. Lehmer has raised the question of the behavior of $\min_n \max |a_j|$, the minimum being over all sequences of length n. The paucity of difference sets with small a implies that sequences attaining the minimum will in general not come from difference sets. On the other hand, the condition $a_j \equiv n \pmod 4$ and the non-existence of e.g. a cyclic difference set (16, 6, 2) shows that this minimum is at least 4 for length 16. For small lengths, it is not hard to construct sequences which attain the minimum after the $a \equiv n \pmod 4$ and non-existence of difference sets constraints. Thus, for length 16 any sequence with $|c_j| < 3$ for all j will satisfy $|a_j| \leq 6$; but $a_j \equiv 16 \pmod 4$ will imply $|a_j| \leq 4$. (Below, we give a construction based on the quadratic character modulo 17 which will give another such sequence). Sequences with $|a_j| \leq 4$ are easily constructed for all lengths up to 32 . (If p is prime $\equiv 1 \pmod 4$, the function which is -1 on the quadratic nonresidues, 1 otherwise, has $a_j = -3$ if j is a nonresidue, 1 if j is a residue). It is interesting to notice that while apparently b(n) goes to infinity with n, we do know of an infinite number of n for which $|a_j| \leq 1$ (the primes $\equiv -1 \pmod 4$) and $2^k - 1$ for all k; Gordon, Mills and Welch have shown that the number of inequivalent sequences gotten by their construction goes to infinity for a suitable sequence of k).

The procedure alluded to previously for finding functions with specified autocorrelation functions works out well for functions defined on a finite abelian group. Thus, for difference sets, we must have $|\sum y_g X(g)| = \sqrt{n}$ for y_g = characteristic function of the difference set, i.e. 1 if g is in the set, 0 otherwise; $n = k - \lambda$, X any non-principal character. The expression $\sum y_g X(g) = X(D)$ is an algebraic integer, and the first step is to see whether the corresponding ideal factorization is possible, and if it is to see whether there are algebraic integers which have the indicated factorization into ideals and which might be of the form $\sum y_g X(g)$. This is the method in the papers of Mann, Rankin, Turyn, Yamamoto. The main conclusions of these papers are negative, and the conclusion usually rests on the assertion that a rational integer must divide the coefficients or differences of coefficients of any algebraic integer which might be X(D), and that this rational integer is too large. Also, relations which must hold between the X(D) put constraints on what algebraic integers the X(D) might be. We shall give several applications of this method below.

Example: The projective plane of order 18. $v = 7^3$, $n = 18$ (see Roth). The question is whether there is a non-commutative group which contains a difference set with these parameters. A group of order p^3 which is not cyclic has a homomorphism onto $Z_p \times Z_p$, and we can apply the same calculation as in the case of difference sets, except that now the y_g lie between 0 and 7 (they represent the number of elements in D in a coset of the kernel). We have $k = 19$, $|X(D)| = \sqrt{n} = 3\sqrt{2}$ for any X . In the field of 7th roots of 1, 3 remains prime, and $2 = \eta\bar{\eta}$, $\eta = \zeta + \zeta^2 + \zeta^4$. Thus (taking $\bar{X}$ instead of X if necessary) $X(D) = \pm 3\eta\zeta^a$. Since $k = 19$, we must have $X(D) = -3\eta\zeta^a$, and then $Y_{a+r} = 1$, $Y_m = 4$ for $m \neq a + r$, with $r = 1, 2$ or 4 . Y_i is the number of elements g of D with $X(g) = \zeta^i$. The factorization of X(D) is a quick and efficient way of finding solutions to the correlation equations which the Y_i must satisfy. (In fact, in this case, the only possible solution for the Y_i was overlooked at first). Now, by Hall's multiplier theorem, 2 is a multiplier (the theorem applies even if the y_g are > 1) and there is a translate of D which is left fixed by the multipliers (i.e. the y_g are left invariant). Thus we may take D so that all the X(D) are -3η or

$-3\bar{\eta}$, $X \neq X_0$. Now by the inversion formula, to find y_e

$$y_e = \frac{1}{7^2}\left(\sum_X X(D)\,\bar{X}(e)\right) = \frac{1}{7^2}\left(19 + \frac{342}{2}\,3\right)$$

since there are $\frac{342}{2}$ pairs, $X, \bar{X}$, $X \neq X_0$, and for each pair $X(D) + \bar{X}(D) = -3(\eta + \bar{\eta}) = 3$. Thus y_e is not an integer, and so D cannot exist. (a is not small here, but this is an interesting example).

Theorem: There is no cyclic difference set with $v = 4.39^2$, $n = 39^2$ $(a = 0)$.

$3^3 \equiv 1 \pmod{13}$, and in the field of 13th and 169th roots of 1, 3 is a product of 4 prime ideals. In the field of 13th roots of 1, let $\eta_0 = \zeta + \zeta^3 + \zeta^9$, $\eta_i = \sigma^i(\eta_0)$ where σ is the automorphism defined by $\sigma(\zeta) = \zeta^2$. The η_i are the fourth-power periods, and form an integral basis for the field they generate. It is easy to verify that $\pi\,\eta_i = 3$, so that the (η_i) are the four prime ideals (3) splits into. Up to an automorphism of the field and a root of 1 factor, the possible factorizations of an integer A such that $A\bar{A} = 3^2$ are $(\eta_0\eta_1)^2$, $\eta_0\eta_1^2\eta_2$, 3; these are in fact integers of absolute value 3 . The first of these (using $\eta_0\eta_1 = \eta_0 + \eta_1 + \eta_3 = -1 - \eta_2$) is $(1+\eta_2)^2 = 1 + 2\,\eta_0 + 2\,\eta_2 + \eta_3$. The second is $1 + \eta_1 + 2\eta_2 + 2\eta_3$.

Now suppose we have an integer B of absolute value 39 in the field of 169th roots of 1 . Then 13 divides B, and $A = B/13$ is an algebraic integer of absolute value 3 . Now if A_1 and A_2 are two such integers, such that $A_1 + A_2$ is divisible by 2, each of A_1, A_2 is of one of the above forms and it is clear that because the η_i are an integral basis, one of $1/2(A_1 \pm A_2)$ will have an expression as a root of 1, times a linear combination of the 1 and three of the η_i with at least one coefficient zero and at least one coefficient of absolute value ≥ 2 . (In fact, $A_1 = \pm A_2$ unless one is of the first form and the other a complex conjugate of one of the second form, times a root of 1 factor). Now apply this to $X(D)$ and $X_2X(D)$, where X is a character of order 169, X_2 a character of order 2 . Then one of $1/2\,(X(D) \pm X_2X(D))$ must be of the above form, since $X(D) + X_2X(D) = 2\sum X(g)$, summed over those g for which $X_2(g) = 1$. We conclude that we have a sum of the form $\sum Y_i\zeta^i$, where $0 \leq Y_i \leq 18$, with ζ a 169th root of 1, and the sum has a coefficient of absolute value ≥ 26, when it is multiplied by a root of 1 and expressed in terms of 1 and three η_i . But this is impossible, since $|Y_i - Y_j| \leq 18$. Thus there is no cyclic $v = 4n$, $n = 39^2$, difference set (in fact none in any group which has a character of order 2.13^2 .

Since there is no known method for constructing binary sequences of length n with $\text{Max}|c_j| = b(n)$ (and no knowledge of the behavior of b(n)) attempts have been made to find sequences with relatively small $\text{Max}|c_j|$ by starting with sequences for which a_j is small and seeing what $\text{Max}|c_j|$ turns out to be. There are no large classes of sequences known with small a_j, except for the Hadamard difference sets (and the quadratic residues for primes $\equiv 1 \pmod 4$). With a sequence for which the a_j are known small, there is a chance that the c_j are small; one has to make the sequence non-periodic and check $\text{Max}|c_j|$. Empirically, it seems that the quadratic residues yield the sequences with smallest $\text{Max}|c_j|$, and the sequence which does is the sequence $X(i)$ rotated by approximately one quarter of the period, i.e. $X(i + \frac{p+1}{4}) = x_i$ (X the quadratic character modulo the prime p). It may be possible to prove that this sequence will have $\text{Max}|c_j|$ very close to minimum in the class of sequences $(X(i+a))$, (perhaps with $X(0) = 0$, which is a more elegant definition). Empirically, $\text{Max}|c_j|$ varies a great deal more with a than for the other Hadamard sets (perhaps because there are more multipliers).

3. Non-binary periodic sequences

We shall now consider periodic sequences whose terms are m-th roots of 1, for $m > 2$, and some related constructions for $m = 2$. If m is a prime power, $m = p^a$, ζ a primitive m-th root of 1, a sum of powers of ζ is zero only when the number of terms is divisible by p, since it must be $\equiv 0 (\text{mod}(1-\zeta))$. A natural question to ask is whether there exist sequences with $a_j = 0$, $j \neq 0$.

Theorem: If q is an odd prime power, there exists a sequence of length q, with terms q-th roots of 1, with $a_j = 0$. For any n, there exists a sequence of length n^2 with terms n^{th} roots of 1, $a_j = 0$.

For the first part of the theorem, use the sequence (x_i), with $x_i = \zeta^{i^2}$, ζ a primitive root of 1. For the second part, we have the sequence obtained by enumerating the rows of the matrix $[\zeta^{ij}]$, ζ a primitive n-th root of 1, considered by Frank and Heimiller. The last is an interesting example of changing one abelian group for another: $[\zeta^{ij}]$ as a function on the group $Z_n \times Z_n$ has auto-correlation function 0 (this is the statement of the orthogonality of the group character matrix for this group). Yates has shown the rows and columns can be arbitrarily permuted for n prime.

We conclude that for any odd n, there is a cyclic Hadamard matrix ($HH^* = nI$) of order n, with n-th roots of 1, i.e., a sequence with $a_j = 0$ of length n. It seems likely the statement is not true for $n = 2p$, p prime. For example, for $n = 6$, we would have to have, for such a sequence x_i, $|\sum x_i \zeta^{ki}| = \sqrt{6}$, with ζ a sixth root of 1, which cannot happen as 2 does not factor in the field of sixth roots of 1.

In general, the method of factoring the Fourier transform works in looking for sequences with specified correlation functions, but there is a new complication, in that more equations may have to be examined. $\sum x_i\zeta^i$ and $\sum x_i \zeta^{ki}$ are not conjugate necessarily algebraic integers, if k is relatively prime to the order of ζ; if the order of the x_i is $>$ and has a factor in common with the length, i.e., the order of ζ. As an example, we consider the next possibility for a cyclic Hadamard matrix with cube roots of 1 as entries.

Theorem: There is no sequence of length 12 composed of cube roots of 1 with $a_j = 0$ for all j.

Proof: Such a sequence, say (ω_j), would satisfy $\sum\omega_j(\omega i)^{kj} = 2(1-\omega)w_k$ with w_k a root of 1, since (2) and (3) factor as $(1-i)^2$ and $(1-\omega)^2$ respectively in the field of 12th roots of 1. By taking complex conjugates if necessary, and multiplying by ω or ω^2 (note that $\omega - 1 = \omega(1-\omega^2)$ to get rid of a -1 factor) we may assume $\sum\omega j = 2(1-\omega)$, i.e., $w_0 = 1$. By rotating the sequence suitably, we may assume $w_3 = i\,\omega^a$, so that then $\omega_9 = -i\omega^a$, as the sum $\sum\omega_j(i\omega)^{kj}$ is $\sigma(\sum\omega_j(-i)^{kj})$ with σ the automorphism defined by $\sigma(\omega) = \omega$, $\sigma(i) = -i$.

Let $X_\ell = \sum_{j=0}^{2} \omega_{\ell+4j}$, $0 \le \ell \le 3$.

We have $\sum X_k(-i)^{km} = 2(1-\omega)\,w_{3m}$

and we have arranged $w_3 + w_9 = 0$. Then $4X_0 = 2(1-\omega)(1+w_6)$, so that $2 | 1+w_6$ and $w_6 = \pm 1$, i.e. does not have a non-trivial cube root of 1 factor. (Compare this with e.g. theorem 5 of Turyn, (1), which asserts that under certain circumstances, if D is a difference set and the X(D) all factor the same way, then the X(D) can be assumed equal.)

We now turn to the sums $\sum_k \omega_{j+3k}$ which we again denote by X_j. Thus $X_0 = \omega_0 + \omega_3 + \omega_6 + \omega_9$, and $X_i \equiv 1 \pmod{(1-\omega)}$. w_4 is a sixth root of 1, since $w_4 \in Q(\omega)$, and by rotating the sequence by a multiple of 4 (thus not affecting w_0, w_3, w_6, w_9) we can assume $w_4 = \pm 1$.

Then $3X_0 = 2(1-\omega)(1+w_4+w_8)$.

This shows $w_4 = -1$ is impossible, since then we would have $1-\omega | 2w$ with w_8 a root of 1. But now, since we changed w_4 only by a factor ω^a, we see that w_8 must be of the form ω^b (since we might have changed it instead by ω^{-b} and concluded that the changed w_8 is +1, not -1). $b = 0$, i.e. $w_8 = 1$ is impossible, since we would then have $X_0 = 2(1-\omega)$, whereas $X_0 \equiv 1 \pmod{(1-\omega)}$. We therefore have two cases:

Case 1: $w_8 = \omega$, $3X_0 = 2(1-\omega)(2+\omega) = 6$, $X_0 = 2$.

This is impossible, since we must have $X_0 \equiv 1 \pmod{(1-\omega)}$.

Case 2: $w_8 = \omega^2$, $3X_0 = 2(1-\omega)^2$, $X_0 = -2\omega$.

In this case, X_0 must be $1+1+\omega^2+\omega^2$ (in some order). We also have $X_1 = -2\omega$, $X_2 = -2\omega^2$, so $X_2 = 1+1+\omega+\omega$.

We will now let $Y_j = \sum \omega_k (-1)^k$, sum over $k \equiv j \pmod 3$, i.e., $Y_0 = \omega_0 - \omega_3 + \omega_6 - \omega_9$, etc. Then, from the expressions derived above for the X_i, we see that Y_0 and Y_1 can only be 0 or $\pm 2(1-\omega^2)$, and Y_2 can be 0 or $\pm 2(1-\omega)$. But we also have

$$3Y_0 = 2(1-\omega)(w_6 + w_{10} + w_2)$$

$$3Y_1 = 2(1-\omega)(w_6 + w_{10}\omega^2 + w_2\omega)$$

$$3Y_2 = 2(1-\omega)(w_6 + w_{10}\omega + w_2\omega^2) \quad .$$

All 3 of the Y_i cannot be 0, as this would imply $w_2, w_6, w_{10} = 0$. Since $|w_i| = 1$, if $Y_0 \neq 0$ we would have $\pm 3\omega^2 = w_6 + w_{10} + w_2$, so $w_6 = w_{10} = w_2 = \pm\omega^2$. But since $w_6 = \pm 1$, this is impossible. Similarly, $Y_1 \neq 0$ is impossible. Thus we must have $Y_0 = Y_1 = 0$, $Y_2 = \pm 2(1-\omega)$, and the $\pm$ sign must equal w_6. $w_{10} = \omega^2 w_6$, $w_2 = \omega w_6$.

Now $12\omega_0 = 2(1-\omega)(\sum w_i)$, so that $\sum w_i = 2(1-\omega^2)\omega_0$. We know that $2 \mid w_i + w_{i+6}$ for $i = 1, 3, 5$, since then $w_{i+6} = \pm w_i$, and $w_2 + w_6 + w_{10} = 0$. Therefore, since $2 \mid \sum w_j$, we must have $2 \mid (w_0 + w_4 + w_8) = 2 + \omega^2$, a contradiction. Thus there are no sequences with cube roots of 1 of length 12 with 0 correlation.

Of course, the Kronecker product construction can be used to give one with sixth roots as entries $(1,1,1,-1$ on Z_4, $1,\omega,\omega$ on $Z_3)$. The Kronecker product here is the usual definition: we have functions f, g defined on groups A, B and we define $f \cdot g = h$ on the direct product $A \times B$ by

$$h(a,b) = f(a)\,g(b) \quad .$$

When m, n are relatively prime, we have $Z_m \times Z_n$ isomorphic to Z_{mn}. The tensor product of two sequences as defined before corresponds to enumerating the elements of the matrix (a,b) in order: row 1, row 2, etc. rather than along the main diagonal, (periodic with respect to both sides) which gives the isomorphism of $Z_m \times Z_n$ with Z_{mn}.

We can also prove, by analogy with the binary case, that the previously exhibited sequences are somewhat unique.

Theorem: Let q be a prime power, (x_i) a sequence of q-th roots of 1 of length q^m with $a_j = 0$ for all j. Then $m \leq 2$.

Proof: Assume the contrary, and let ζ be a primitive q^m-th root of 1. Then we must have $|\sum_i x_i \zeta^i| = q^{m/2}$. q factors as a power of a single ideal in $Q(\zeta)$ and there is already an integer A of absolute value $\sqrt{q}$ in the field of q-th roots of 1, say $Q(\zeta^r)$, $r = q^{m-1}$; thus, after rotating the x_i to remove a root of 1 factor if necessary, we may assume that $\sum x_i \zeta^i = B$, with $|B|^2 = q^m$, $B \in Q(\zeta^r)$, and since ζ is of degree r over $Q(\zeta^r)$, if $B_i = \sum_0^{q-1} x_{i+jr}$, $B_0 = B$, $B_i = 0$ for $i > 0$. But then $q^{m/2} = |B| = |B_0| \leq q$, so $m \leq 2$. (The proof has to be changed slightly for $q = 2$).

It can easily be shown that a sequence of order p^2 of p-th roots of 1 with $a_j = 0$ must be one of the sequences obtained by rearranging the rows and columns of $[\zeta^{ij}]$ and enumerating the rows.

The proof of the preceding theorem can be generalized, by analogy with the theorems which apply when the terms of the sequence are rational (0 and 1 or ± 1) to prove the non-existence of various cyclic Hadamard matrices with entries m^{th} roots of 1, $m > 1$. In the binary case, it is known (Turyn, (1), theorem 8) that if n and v are both even there is no cyclic Hadamard matrix (of course, v has to be even). There are no cyclic difference sets known in which n and v have a factor in common, and probably none such exist, at least if the prime divides v^2 to a higher power than n. If we look for a cyclic matrix with p^m-th roots of 1 as entries, the length has to be a multiple of p; there are probably some deep non-existence theorems about such matrices.

A cyclic Hadamard matrix with roots of 1 as entries satisfies

$$HH^* = vI, \quad HJ = JH = cJ$$

for some constant c, the same equations that the incidence matrix of a (v, k, λ) design satisfies (even when ± 1 entries are used) and matrices which satisfy

$$MM^* = vI + aJ, \quad MJ = M^*J = cJ$$

for suitable v, a, c are interesting ("multi-valued logic") generalizations of v, k, λ designs. Cyclic matrices of this type correspond to sequences with constant a_j. Zierler has constructed such sequences of length $p^n - 1$ with $a_j = -1$ (they arise from the Singer construction as in the binary case) with p^{th} roots of 1 as elements; it is interesting to note that they are of the form $A \cdot B$, where B is the sequence $(1, \zeta, \ldots, \zeta^{p-2})$. Butson has constructed Hadamard matrices of order $2p$, with p^{th} roots as entries, and the core of such a matrix satisfies the conditions above.

It can be shown that there are no sequences of cube roots of 1 of lengths 11, 14, 17, 20, 23, 29, 38, 41 with $a_j = -1$ for all j, by methods similar to the above. The case 23 is of particular interest: it can be shown that any sequence of 24th roots of 1 with $a_j = -1$ must be equivalent to the quadratic residue sequence, so that cube roots are impossible. This is of interest because Hall has shown that the Mathieu group M_{12} is the automorphism group of the binary Hadamard matrix of order 12; no similar representation is possible for M_{24}.

Another example of a sequence with small constant a_j is the sequence of length p, p prime $\equiv 1 \pmod 4$ defined by $x_0 = i$, $x_j = X(j)$, X the quadratic character mod p. That is an immediate consequence of the next theorem, which is a summary of known facts. It might be mentioned that an Hadamard matrix with fourth roots of 1 entries, will give a binary Hadamard matrix if i^a is replaced by $T^a \times H$, H binary Hadamard, T a monomial matrix which satisfies $T' = -T$, $T^2 = -I$, e.g. $\begin{pmatrix} 0 & 1 \\ -1 & 0 \end{pmatrix}$ repeated on the main diagonal. There is no such representation possible for $\zeta^p = 1$, p odd prime.

This is the crux of one of Williamson's constructions, using the preceding construction for an arbitrary $q \equiv 1 \pmod 4$.

Theorem: Let q be a prime power, e a divisor of $q-1$, and X a character of order e on GF(q) (X multiplicative, $X(0) = 0$, $X^e(x) = 1$ for $x \neq 0$). If M is the matrix $[X(a_i - a_j)]$ (the a_i elements of GF(q); in any order) then M is 0 only on the main diagonal, and

$$MM^* = qI - J$$

$$M' = X(-1)M .$$

Write $M = \sum_0^{e-1} \zeta^i M_i$, with ζ a primitive e^{th} root of 1, M_i matrices with 0,1 entries. Then

$$JM_i = M_i J = \frac{q-1}{e} J$$

$$[M_i M_j']_{rs} = (j-m,\ i-m)$$

((a,b) the cyclotomic number, $X(a_s - a_r) = \zeta^m$) .

We prove only the first assertion. $\sum X(z) = 0$ as z ranges over GF(q); thus $\sum X(z)\bar{X}(z+\alpha) = \sum_{z\neq 0} \bar{X}(1+\alpha z^{-1})$, so the sum is $0 - \bar{X}(1) = -1$, as $1 + \alpha z^{-1}$ will range over all $y \in GF(q)$ except $y = 1$. (This proof is due to Williamson, for $e = 2$). The rest of the theorem is equally straightforward.

These matrices, with X the quadratic character, have been used in the construction of Hadamard matrices. In fact, almost all constructions known depend ultimately on such matrices, and most of these constructions do not use any particular arrangement of the a_i. For $q \equiv -1 \pmod 4$, the matrix with identity added is the core of an Hadamard matrix. Such matrices have been generalized to the skew

Hadamard matrices (core $-I$ is skew). For example, the Stanton-Sprott construction of a difference set of order q_1q_2 yields immediately the Ehling construction when the fields $GF(q_i)$ are replaced merely by matrices M which satisfy the equations above. Of course, it is natural to make the matrix cyclic if q is a prime, or in general the group matrix corresponding to the additive group of $GF(q)$. However, there is another natural way to enumerate $GF(q)$: $0, 1, g, g^2, \ldots$ where g is a primitive element of $GF(q)$. With this arrangement of the elements of $GF(q)$, the matrix M becomes (omitting the symbol X)

$$\begin{matrix} 0 & 1 & g & g^2 & g^3 & \cdots \\ -1 & 0 & g-1 & g^2-1 & g^3-1 & \cdots \\ -g & 1-g & 0 & g(g-1) & g(g^2-1) & \cdots \\ -g^2 & 1-g^2 & g(1-g) & 0 & g^2(g-1) & \cdots \end{matrix}$$

Aside from the properties mentioned in the theorem above, a description of the matrix is as follows: if $X(g) = \zeta$, the first row is $1, \zeta, \zeta^2, \ldots$ (if X is the quadratic character, the first row is $0, -1, +1, -1, \ldots$). The first column is identical except for the $X(-1)$ factor. The core of the matrix is almost cyclic: the second row is a translate of the first, but with a $X(g)$ factor. Note that the element "lost" at the end of the second row is $(g^{q-2} - 1)g = 1 - g$, since $g^{q-1} = 1$, which reappears at the beginning.

We now summarize the previous remarks:

Theorem: If q is a prime power, $e \mid q-1$, ζ a primitive e^{th} root of 1, $X(-1) = 1$ if $e \mid \frac{q-1}{2}$, -1 otherwise, then there exist sequences (x_i) of length $n = q-2$, with x_i e^{th} roots of 1, which satisfy

$$x_i = X(-1)\, x_{n+1-i}\, \zeta^{-i}$$

and

$$c_j + \bar{c}_{n+1-j} = -1 - \zeta^{-j} .$$

They can be obtained as $x_i = X(g^i - 1)$, with g a primitive element of $GF(q)$.

Also, if $n = 2m + 1$, $e = 2$, $u = X(-1) = (-1)^{m+1}$

$$\sum_1^{m-k} x_i x_{i+2k} + \sum_1^{k-1} x_i x_{2k-i} (-1)^i u = -1 + (-1)^{k+1} \frac{(1+u)}{2} \tag{5}$$

$$\sum x_i = \pm 1, \quad (\sum x_i)(\sum x_i (-1)^i) = u \tag{6}$$

$$|\sum x_i \zeta^i| = \sqrt{q} \text{ if } \zeta^{q-1} = 1, \ \zeta^2 \neq 1 . \tag{7}$$

Equation (5) above is a compression, using the reflection rule, of the dot product rule, the second equation in (6) is the dot product with the first row statement, and (7) is a simple computation. Note the analogy, in (7) to a difference set: in a difference set all the Fourier transforms except the one corresponding to the principal character have the same absolute value. Here we exclude the subgroup of the character of order 2; this is like the Gordon - Mills - Welch condition on the polynomial $\Omega(x)$ which is the "quotient" of two corresponding Singer sequences.

For $n \leq 21$, $(e = 2)$ all the sequences which satisfy the conditions of the theorem arise from finite fields in the manner described, so that in particular there are none for $n = 13$. Note that for $e = 2$, the reflection rule in the above theorem is the same as for the Barker sequences, (4). In fact, the Barker sequences of lengths $3, 7, 11$, which we identified with difference sets corresponding to these primes, arise from the fields of $5, 9, 13$ elements respectively as described in the theorem. (There are in general $1/2\ \phi(q-1)$ such sequences, if we do not differentiate between sequences generated by g and g^{-1}, which are the same sequences but written backwards. For $q = 13$, $g = 2$ does not yield a Barker sequence but $g = 6$ does. Again, for $q = 13$, $e = 3, 4, 6$ do not yield sequences with $|c_j| \leq 1$). Note the second equation of the theorem, which has an expression analogous to a_j, but is not a_j: $a_j = c_j + \bar{c}_{n-j}$. The expression is the a_j of the sequence of length $n+1$ with an initial 0 adjoined: if 0 is replaced by 1, the resulting sequence of length $q-1 = n+1$ satisfies $a_j = -1 - \zeta^{-j} + \bar{x}_j + x_{n+1-j}$. Thus $|a_j| \leq 4$ for all j; such sequences $(e = 2)$ were mentioned above. The Barker sequences of lengths 5 and 13 do not satisfy the conditions of the theorem (they do not even satisfy $\sum x_i = \pm 1$).

It can be shown, by methods depending on factorization of the sums $\sum x_i \zeta^i$, that there are no such binary sequences of length $q-2$ if q is not a prime power <225. For most odd numbers <225 which are not prime powers, it is easy to find a prime divisor of q and a divisor of $q-1$ to show that there is no possible factorization into ideals for a number of absolute value q in all fields of $(q-1)$ roots of 1. For example, if $4 | q-1$ and q contains a prime $\equiv -1 \pmod 4$ to an odd power, there is no integer of absolute value $\sqrt{q}$ in $Q(i)$, e.g. for $q = 21$. Some numbers q are more difficult: for these a more careful examination must be made, and an argument either on the size or parity of the coefficients can be made. Some examples follow:

$q = 45 = 9.5$, $q - 1 = 4.11$. Here we denote the sequence by x_i, and let $X_i = \sum x_{i+11j}$. Thus $X_0 = x_{11} + x_{22} + x_{33}$ is odd, the others even. We have the factorizations $3 = \eta\bar{\eta}$, $5 = (2+\eta)(2+\bar{\eta})$ $\eta = \sum \zeta^r$, r a residue mod 11, ζ a primitive 11th root of 1, so that, up to an automorphism and a root of 1 factor, the integers of absolute value $\sqrt{45}$ in $Q(\zeta)$ are $4\eta + 3$, $7 + \eta$ and $3(2+\eta)$. It is clear that the second and third cases are impossible because of the parity of the X_i, and in the first case the root of 1 factor must be $w = \pm 1$, and we have

$$X_r - X_2 = 4w ,$$

$$X_0 - X_2 = 3w$$

$$X_n - X_2 = 0$$

r, n arbitrary residues and nonresidues mod 11, respectively. Assuming the X_i normalized so that $\sum X_i = 1$, we have

$$1 = 3w + X_2 + 5(4w + X_2) + 5X_2$$

$$1 = 11X_2 + 23w$$

and thus $w = +1$, $X_2 = -2$, $X_n = -2$, $X_r = 2$, $X_0 = 1$. Since $x_i = -x$ for i odd, $X_0 = 1$ implies $x_{22} = 1$. We now let $Z_i = \sum x_{i+11j}(-1)^{i+11j}$; as before, (using now the sum with a primitive 22nd root of 1) we conclude that $Z_0 = 1$, and $Z_r = -Z_n = \pm 2$ (since the sum could be $3 + 4\bar{\eta}$). In case $Z_r = 2$ we conclude that $X_i + X_{22+i} =$ for i an even residue e.g. (12, 14, 4, 16), and $x_n + x_{22+n} = -2$ for n an even nonresidue (2, 6, 18, 8, 10). But then $x_{12} = x_{34} = x_{14} = x_{36} =$ (etc) $= 1$, $x_2 = x_{24} = x_6 = x_{28} =$ (etc) $= -1$. Now when we form $\sum_m x_m i^m$, the terms $x_m i^m + x_{m+22} i^{m+22} = (x_m - x_{m+22}) i^m$ will cancel for all even m, and the sum will have to be $1 + a$ pure imaginary number and of absolute value $\sqrt{45}$, which is impossible, as such a number is $\pm 6 \pm 3i$ or $\pm 6i \pm 3$. In case $Z_r = -2$, $Z_n = +2$ we conclude as before that $x_i + x_{i+22} = 2$ for i an odd nonresidue, so again $x_i = x_{i+22}$ for i odd, and the sum would be a real integer of absolute value $\sqrt{45}$.

$q = 99$, $q - 1 = 2.7^2$. 3 is prime in the field of 49th roots of 1, so that $3 | \sum x_i \zeta^i$ for $\zeta^2 \neq 1$. Now in the inversion formula $x_j = \frac{1}{98} \sum_k (\sum_i x_i \zeta^{ki}) \zeta^{-kj}$, there is no 3 in the denominator so that $3 | x_j$ whenever $\sum x_i$ and $(\sum x_i (-1)^i) (-1)^j$ cancel, which is half the time.

$q = 175$, $q - 1 = 2.3.29$. It is easy to verify that $5^7 \equiv -1$ (mod 174), so again we would have 5 dividing half of the x_i.

$q = 119$: here $q = 7.17$, $q - 1 = 2.59$. Mod 59 both 7 and 17 generate all the quadratic residues, so that $\sum x_i \zeta^i = w \sum x_i \zeta^{7i}$, with w a root of 1, ζ a 59th (or 118th) root of 1, $w = \pm \zeta^a$. By a rotation of the x_i, (which would change the equations) we could assume $a = 0$, and we would then have to have $w = 1$, as $\sqrt{-7.17}$ does not belong to $Q(\zeta)$. Now we have an integer of absolute value $\sqrt{109}$ in $Q(\sqrt{-59})$. $2+\eta$, with η a quadratic period, is an integer of absolute value $\sqrt{17}$ in this field, so the quotient of $\sum x_i \zeta^i$ by $2+\eta$ or $2+\bar{\eta}$ would be an integer of absolute value $\sqrt{7}$ in this field. But $|a+b\eta|^2 = a^2 + 15b^2 - ab = (a-\frac{b}{2})^2 + \frac{59b^2}{4}$ and it is easy to see that we cannot have this equal to 7, i.e., there is no integer of absolute value $\sqrt{7}$ in $Q(\eta)$.

$q = 153$: $q = 9.17$, $q-1 = 8.19$. Since $3^9 \equiv -1 \pmod{4.19}$, we have $3 | \sum x_i \zeta^i$ for $\zeta^2 \neq 1$, $\zeta^{76} = 1$. Let $X_i = x_i + x_{i+76}$. Again, $3 | X_i$ for i odd, so $x_i = -x_{i+76}$ for i odd, and $3 | X_i - X_0$ for i even, so $x_i = x_{i+76} = -x_{76}$ for i even. Thus we cannot have $\sum x_i = 1$, as approximately 3/4 of the x_i are $= x_{76}$.

4. Miscellany

The most interesting question which can be raised is the behavior of $b(n,m)$ for large n, where the m indicates the order of the roots of 1 considered, $b(n,m) = \min_n \max_j |c_j|$. Golomb and Scholtz have considered Barker sequences for $m \leq 6$. There might be a much better opportunity for Barker sequences for $m > 6$, even m prime, because there are more types of integers of absolute value <1. For example, for $m = 7$, $1 + \zeta^2 + \zeta^4$ and $1 + \zeta + \zeta^4$ both have absolute value <1, and thus there are two different integers $\equiv 3 \pmod{(1-\zeta)}$ with absolute value less than 1. For $m = 3, 4$, the Barker condition is quite severe: we have $a_j = c_j + \bar{c}_{n-j} = \sum x_i \bar{x}_{i+j}$, and $c_j \equiv X(n-k) \pmod{(1-\zeta)}$ so that, since the sum of the exponents in $a_j \equiv 0$ mod 3 or mod 4, respectively, a_j is real or $\pm 2i$. Thus for $m = 3$, n a multiple of 3, a Barker sequence leads to a cyclic Hadamard matrix, and thus cannot exist for n divisible exactly by an odd power of a prime $\equiv -1 \pmod 3$, e.g. $n = 15, 33$. There are probably no cubic Barker sequences for $n > 9$. For $m = 4$, we get similar restrictions: we must have a_j real if n is odd, so $c_j = \pm i$ is impossible, since $c_j + \bar{c}_{n-j}$ must be real and $c_j = 0$ for $n-j$ even. The Frank sequences (Turyn, (2)) do not seem to give a marked improvement as n increases: $\max|c_j|$ is approximately $\frac{\sqrt{n}}{\pi}$ no matter what the "alphabet size" is.

The tables at the end, which show $\max|c_j|$ for sequences $X_3(g^i-1)$, $X_4(g^i-1)$, $X_3(i+a)$, $X_4(i+a)$ for various p do not indicate any large advantage over the binary sequences. The results of Boehmer also seem to imply that there is no power residue definition, modulo a prime, which is likely to be better the sequences $X_2(i+a)$.

There are two other generalizations of some interest: different abelian groups, and more than one sequence. There are difference sets with $a = -1$ in other abelian groups, e.g. the squares in $GF(q)$ (all the cyclic constructions mentioned for p prime generalize to

GF(q)), and the twin-primes construction. (Question: Does the Singer construction generalize to other abelian groups?) There are several classes of difference sets with $a = 0$ in other abelian groups, but the cyclic subgroups are very small (Turyn, (1)). There is the Kronecker product construction for these, and the basic building blocks are:

1) Any one element set in a group of order 4.
2) The set of all vectors with Hamming weight of the form $4m$ or $4m+1$ in the abelian group of type $(2,2,\ldots,2)$ (an even number of copies).
3) The set in 2) can be changed to the following: the group is as in 2), and the set is all vectors with an odd number of blocks of 1's an d first component $=1$, or an even number of blocks of 1's, first component $= 0$.
4) The subset of all $(a+b, ab)$ of $GF(2^h)\times GF(2^h)$.
5) G is of type $(4,3,3)$ or $(2,2,3,3)$: we pick four lines of different slopes in $GF(3)\times GF(3)$, L_i; here $0,1,2,3$, is an enumeration of the group of order 4 which occurs in G, $D = (0,\tilde{L}_0) \cup (i, L_i)$, $i \neq 0$.

Thus there are such sets in the abelian group $Z_3 \times Z_{12}$ and therefore in $Z_{12}\times Z_{12}$ (taking product with Z_4), but the known constructions do not yield a cyclic subgroup of order > 24 $(Z_8 \times Z_2 \times Z_3 \times Z_{12})$, though there undoubtedly are some.

Welti and Golay have independently studied the possibility of finding two binary sequences, say X and Y, such that $c_j(X) + c_j(Y) = 0$ for $j \neq 0$ (see also Turyn (3)). The basic theorem is easy to generalize to sequences of nth roots of 1: if we have n sequences X_i such that $\sum_i c_j(X_i) = 0$ for all n, then the n sequences

$$(X_1,\ldots,X_n)\,[\zeta^{ij}]$$

also have this property, i.e. the sequences

$$X_1, X_2, X_3, \ldots$$
$$X_1, \zeta X_2, \zeta^2 X_3, \ldots$$

etc. Any n one-term sequences have this property. It follows immediately, of course that $\sum_i c_j(X_i) = 0$ implies $\sum_i a_j(X_i) = 0$.

Quadruples of sequences such that $\sum_{i=1}^{4} a_j(X_i) = 0$ have been investigated: these are called Hadamard matrices of Williamson type (it is also required that $x_i = x_{-i}$ in each sequence). (see e.g. Hall's book). These do not necessarily correspond to difference sets in a quaternion group. If we have a group with a normal subgroup, say cyclic and of index 4, and a subset in the group, we can associate four sequences with the subset, the incidence sequence for each coset. The condition that the subset be a difference set becomes a condition on the autocorrelation functions of the four sequences and on the cross-correlations. Precisely which cross-correlations depends of course on the multiplication in the group. The sets with $a = 0$ in the groups of order 36 are of this "switching group" type, i.e. one definition works for several groups. Here the normal subgroup is $Z_3 \times Z_3$, not cyclic.

Table 1

Sequences with small $\max_j |c_j|$

n	Sequence	$\max \lvert c_j \rvert$
14	5 2 2 2 1 1 1	2
15	5 2 2 1 1 1 2 1	2
	6 2 2 1 1 1 2	2
16	3 1 3 4 1 1 2 1	2
	5 2 2 2 1 1 1 2	2
17	2 2 5 1 1 1 1 2 1 1	2
	4 2 2 1 2 1 1 1 1 2	2
18	5 1 1 2 1 1 3 2 2	2
19	4 3 3 1 3 1 2 1 1	2
20	5 1 3 3 1 1 2 1 1 2	2
21	6 1 1 3 1 1 2 3 2 1	2
	5 1 1 3 1 1 2 3 2 2	2
	3 5 1 3 1 2 1 1 1 2 1	2
22	8 3 2 1 2 2 1 1 1 1	3
23	1 2 2 2 2 1 1 1 5 4 1 1	3
24	8 3 2 1 1 1 1 2 2 1 2	3
25	3 2 3 6 1 1 1 1 1 2 1 2 1	2
26	8 2 1 1 2 2 1 1 1 1 1 2 3 1	3
27	2 1 2 1 1 2 1 3 1 3 1 3 4 2	3
28	2 1 2 1 1 2 1 3 1 3 1 3 4 3	2
	3 2 3 6 1 1 1 1 1 2 1 2 1 2 1	2
29	2 1 2 1 1 2 1 3 1 3 1 3 4 4	3
30	7 1 2 2 1 1 1 1 2 1 1 3 4 2 1	3
31	3 2 2 3 6 1 1 1 1 1 1 2 1 2 1 3	3
32	6 1 3 2 1 1 2 1 2 1 1 3 1 1 3 1	3
	6 1 3 2 1 3 3 1 1 2 1 2 1 1 3 1	3
33	6 3 1 2 3 2 1 1 3 2 1 1 2 2 1 1 1	3
	6 3 1 2 2 2 1 1 1 3 2 4 1 1 1 2	3
	6 3 1 2 1 1 1 2 1 2 2 3 1 1 1 3 1 1	3
34	7 4 2 1 1 2 1 1 2 2 2 2 1 1 1 1 1 1 1	3

Table 2

Nonexistence of cyclic difference sets with $a = 0$, $v = 4N^2$, $n = N^2$.

We refer to the following two theorems (Turyn, (1)):

I. We have a character X_1 of order v_1, m a divisor of n such that $m | X_1 X(D)$ for all characters X in a group H of order v_2, $X^j \notin H$ for $0 < j < v_1$. Then $2^{r-1} v \geq m v_1 v_2$, with r the number of distinct prime divisors of v_1 . (If $v_1 = 1$, $v \geq m v_2$.)

II. Assume D is a difference set such that $2^b \| v$, $2^{2a} \| n$, $b \geq a > 0$. Then G cannot have a character of order 2^b, and in particular cannot be cyclic. ($2^b \| v$ means $2^b | v$, $2^{-b} v$ is odd.)

We list below most of the restrictions on abelian groups in which we might have a set with $a = 0$. Note that for $N = 2^t$, I gives a stronger restriction than II .

We omit the statement $v_2 = 1$ or $m = N$ where applicable. An entry of e.g. 3 under p and 4 under bound means there can be no character of order 3^5 .

N	Reason	v_1	v_2	m	p	Bound on Exp.
2	II				2	3
3	I	3^2	2		3	1
4	I	2^5			2	4
5	I	5^2	2		5	1
6	II				2	3
7	I	7^2			7	1
8	I	2^6			2	5
9	I	3^4			3	3
10	II				2	3
11	I	11^2			11	1
12	II				2	5
13	I	13^2			13	1
14	II				2	3
15	I	5^2	18	5	5	1
	I	3^2	50	3	3	1
16	I	2^7			2	6
17	I	17^2			17	1

Table 2 Continued

N	Reason	v_1	v_2	m	p	Bound on Exp.
18	II				2	3
	I	3^4	4		3	3
19	I	19^2			19	1
20	II				2	1
21	I	3^2	98	3	3	1
22	II				2	3
	I	11^2	2	11	11	1
23	I	23^2			23	1
24	II				2	7
25	I	5^3	2		5	2
26	II				2	3
	I	13^2			13	1
27	I	3^4	4		3	3
28	II				2	5
29	I	29^2			29	1
30	II				2	3
31	I	31^2			31	1
32	I	2^8			2	7
33	I	11^2	4	11	11	1
34	II				2	3
	I	17^2			17	1
35	I	7^2	50	7	7	1
36	II				2	5
37	I	37^2			37	1
38	II				2	3
	I	19^2			19	1
39						
40	II				2	7
41	I	41^2			41	1

Table 2 Continued

N	Reason	v_1	v_2	m	p	Bound on Exp.
42	II				2	3
43	I	43^2			43	1
44	II				2	5
45	I	3^4	50	9	3	2
46	I	23^2		23	23	1
	II				2	3
47	I	47^2			47	1
48	II				2	9
49	I	7^4			7	2
50	II				2	3
	I	5^4			5	3
51	I	17^2			17	1
52	II				2	5
53	I	53^2			53	1
54	II				2	3
	I	3^5			3	4

Table 3

Nonexistence of sequences analogous to $X(g^i - 1)$ for q not a prime power.

Divisor of q cannot have proper ideal factorization in $Q(\zeta)$, ζ a primitive m^{th} root of 1, $m|q-1$

q	Divisor of q	m
15	3, 5	7
21	3, 7	4
33	3, 11	4
35	5	17
39	3	19
51	3	5
55	5, 11	3
57	3, 19	4

Table 3 Continued

q	Divisor of q	m
63	7	31
69	3, 23	4
75*	5	37
77	7, 11	4
85	5	3
87	3	43
91	7, 13	5
93	3, 31	4
95	5	47
105	3, 7	4
111	3	5
115	5	3
117**	3	29
123	3	61
129	3, 43	4
133	7, 19	4
135	3, 5	67
141	3, 47	4
143	11	71
145	5	3
147***	7	73
155	5, 31	7
159	3	79
161	7, 23	4
165	3, 11	4
171	19	5, 17
177	3, 59	4
183	3, 61	7
185	5	23
187	11, 17	3
189	3, 7	4
195	5	97
201	3, 67	4
203	7	101
205	5, 41	3

* 75: We have $5 \mid \Sigma x_i \zeta^i$ for ζ a 37th, 74th root of 1, so 5 divides x_i for i odd, which is impossible.

** 117: $3 \mid x_i + x_{i+58}$ for i odd, $3 \mid x_i + x_{i+58} - x_0$ for i even, so $x_i = x_{i+58} = -x_0$ for i even, $x_i = -x_{i+58}$ for i odd, which is impossible as before $(\Sigma x_i \neq \pm 1)$.

*** 147: $7 \mid x_i$ for i odd, a contradiction

Table 3 Continued

q	Divisor of q	m
207	23	103
209	11,19	4
213	3, 71	4
215	5, 43	107
217	7, 31	4
221	13,17	11

Goethals and Seidel have investigated such matrices from a different point of view, and have constructed a matrix of order 225 with the right entries and symmetry conditions. However, the matrix is not necessarily of the right circulant form. Hall has also investigated similar matrices.

Table 4

Nonexistence of sets with $a = 3$

p, q divide n and v respectively, and show that the set cannot exist.

s	v	n	k	p	q
5	31	7	10	7	31
6	43	10	15	5	43
9	91	22	36	2	13
10	111	27	45	3	37
13	183	45	78	5	3
14	211	60	91	3	211
17	307	76	136	2	307
18	343	85	153	5	7

Table 5

$\underset{j}{\text{Max}} |c_j|$ for sequences $X_3(g^i-1)$ (c_j must be real)

p	g	$\text{Max}\|c_j\|$	p	g	$\text{Max}\|c_j\|$
13	2	3	67	2	10
	6	3		32	9
19	2	4		61	7
	13	4		18	10

Table 5 Continued

p	g	$\mathrm{Max}\lvert c_j\rvert$	p	g	$\mathrm{Max}\lvert c_j\rvert$
19	14	4	67	20	7
31	3	4		13	7
	17	6		7	8
	13	6		28	7
	24	7		46	7
37	2	4		50	7
	32	4	97	5	12
	17	7		21	10
	13	7		40	12
	15	7		71	10
	18	4		29	13
43	3	7		83	13
	28	7		38	10
	30	7		82	13
	12	7		13	10
	26	6		74	10
	19	4		7	13
61	2	7		10	13
	6	7		56	13
	35	7		80	10
	18	7		60	9
	44	7			
	54	9			
	10	7			
	30	7			
127	3	13			
	116	16			
	109	13			
	92	13			
	86	15			
	12	13			
	83	12			
	112	10			
	55	13			
	114	13	127	67	15
	48	12			
	78	12			
	67	15			
	93	12			
	106	15			
	65	15			
	58	13			
	14	13			

For each p, the sequences corresponding to distinct g have different correlation functions (only one of g and g^{-1} was used).

Table 6

$\text{Max}|c_j|^2$ for sequences $X_3(a+i)$, $a = 0, 1, \ldots, \frac{p+1}{2}$, $X_3(0) = 1$,

$$c_j = \sum_1^{p-1-j} X_3(i)\, \bar{X}_3(i+j)$$

p												
p = 13	9	12	9	4	7	13	21					
p = 19	25	16	16	21	25	21	21	21	19	31		
p = 31	25	28	28	37	19	21	21	28	43	27	27	28
	28	31	43	25								
p = 37	36	36	37	31	21	21	27	43	43	27	21	25
	28	28	37	37	28	43	25					
p = 43	49	36	39	28	27	36	31	52	61	61	79	91
	61	52	37	37	43	37	31	52	49	49		
p = 61	64	67	63	79	79	73	73	73	73	61	75	91
	76	63	63	63	48	48	48	39	57	57	61	84
	81	81	84	63	79	63	57					
p = 67	64	91	64	64	64	67	48	67	64	48	43	37
	39	39	43	37	31	52	61	61	73	67	73	73
	52	49	48	43	64	57	73	73				
p = 97	100	100	81	81	84	76	67	57	75	75	76	109
	133	133	151	139	151	133	127	109	91	109	127	148
	112	112	112	127	127	127	127	148	151	163	163	133
	97	81	76	76	73	91	91	124	103	91	121	124
	157											
p = 127	100	103	100	121	121	157	157	157	133	127	121	124
	163	163	133	133	133	133	108	133	157	196	241	241
	241	241	241	196	157	175	175	139	151	181	151	148
	147	147	139	139	151	133	133	163	208	175	175	133
	133	175	156	193	244	223	183	183	199	163	157	183
	183	157	196	196								

Table 7

$\mathrm{Max}|c_j|$ for $X_2(g^i - 1)$

p	g	$\mathrm{Max}\|c_j\|$	p	g	$\mathrm{Max}\|c_j\|$
11	2	3	97	5	13
	8	3		21	13
13	2	5		40	9
	6	1		71	13
23	5	7		29	13
	10	3		83	13
	20	7		38	13
	17	3		94	13
	11	7		13	17
61	2	9		74	17
	6	9		7	13
	35	13		10	13
	18	13		56	17
	44	9		80	13
	54	9		60	13
	10	9			
	30	9			
67	2	11			
	32	11			
	61	11			
	18	7			
	20	11			
	13	11			
	7	11			
	28	11			
	46	11			
	50	11			

Table 8

$\mathrm{Max}|c_j|^2$ for the sequence $X_4(i+a)$, $a = 0, 1, \ldots, X_4(0) = 1$

$$c_j = \sum_0^{p-1-j} X_4(i)\, \bar{X}_4(i+j) \;,$$

p													
p = 13	10	16	10	17	17	13	9	9	10	10	17	20	29
p = 17	25	25	17	17	17	17	13	16	16	16	16	16	13
	17	17	17	17									
p = 29	17	17	10	20	18	16	26	25	17	17	17	25	20
	18	17	25	25	25	25	25	16	16	16	26	25	17
	20	18	13										
p = 37	37	37	25	25	25	26	34	26	29	40	29	26	41
	37	34	41	41	50	58	40	50	50	26	18	29	29
	29	25	26	37	26	25	41	29	29	29	29		
p = 41	65	65	65	53	37	29	26	40	58	50	37	29	20
	26	29	34	36	40	26	40	36	36	36	40	26	40
	36	34	29	26	20	29	37	50	58	40	26	29	37
	53	65											
p = 53	49	81	65	65	50	50	50	73	65	49	81	65	85
	65	65	53	53	50	65	73	73	73	73	53	49	45
	50	64	50	37	50	50	65	73	73	73	65	50	50
	50	64	64	82	64	82	52	74	80	58	58	53	53
	50												
p = 61	49	64	64	50	41	50	40	58	65	73	53	49	49
	40	40	50	50	64	100	82	64	50	61	41	52	58
	52	37	53	45	49	49	49	37	53	45	45	37	41
	41	41	29	37	53	65	65	53	53	45	52	50	50
	53	68	58	53	65	53	53	37	49				
p = 73	50	50	50	53	65	65	53	73	65	85	61	45	61
	41	52	68	50	58	53	45	68	82	85	85	90	68
	58	50	82	80	68	90	68	90	80	90	82	100	82
	90	80	90	68	90	68	80	82	50	58	68	90	85
	85	82	68	45	53	58	50	68	52	41	61	45	61
	85	65	73	53	65	65	53	50					

Table 8 Continued

p = 89	100	100	104	85	125	109	125	101	65	65	73	73
	89	113	97	85	109	125	125	145	122	104	90	116
	104	116	160	130	116	90	90	90	104	104	80	65
	68	82	82	122	100	144	144	122	144	144	144	122
	144	144	100	122	82	82	68	65	80	104	104	90
	90	90	116	130	160	116	104	116	90	104	122	145
	125	125	109	85	97	113	89	73	73	65	65	101
	125	109	125	85	104							
p = 101	122	122	104	122	148	104	104	90	82	52	61	65
	65	85	85	97	109	125	125	109	73	85	85	109
	101	85	90	73	68	82	82	104	122	104	104	101
	125	121	106	85	81	121	85	85	85	109	122	170
	148	144	144	144	144	144	148	170	122	100	82	65
	68	82	82	90	104	122	122	100	104	104	104	90
	68	85	73	89	85	82	100	104	97	85	73	104
	122	122	104	90	65	65	64	82	61	82	104	122
	104	116	144	122	104							
p = 109	196	170	196	170	178	148	144	144	170	122	104	90
	116	116	116	90	85	109	101	101	101	65	97	109
	121	101	121	125	153	173	145	122	104	101	101	100
	82	82	81	81	85	100	100	137	109	97	97	109
	97	137	137	153	153	145	101	121	101	145	153	153
	137	137	97	101	85	89	113	101	113	89	85	82
	90	82	85	101	100	100	101	121	144	170	148	122
	122	104	122	101	89	85	101	109	109	125	97	85
	80	89	125	97	74	82	122	100	144	122	170	170
	200											
p = 113	169	169	169	173	145	145	173	173	169	145	104	130
	116	106	100	73	80	74	80	82	82	104	122	148
	170	200	200	234	226	170	197	173	125	106	145	146
	145	116	101	101	97	106	146	130	116	106	106	106
	125	122	104	104	122	169	169	169	197	225	197	169
	169	169	122	104	104	122	125	106	106	106	116	130
	146	106	97	101	101	116	145	146	145	106	125	173
	197	170	226	234	200	200	170	148	122	104	82	82
	80	74	80	73	100	106	116	130	104	145	169	173
	173	145	145	173	169							

Table 9

$\text{Max}|c_j|^2$ for sequences $X_4(g^i-1)$

p	g	$\text{Max}\|c_j\|^2$	p	g	$\text{Max}\|c_j\|^2$
13	2	9	137	3	121
	6	9		27	128
61	2	50		106	200
	6	81		132	121
	35	50		92	242
	18	25		6	200
	44	72		54	162
	54	49		47	200
	10	121		108	200
	30	81		13	128
97	5	128		94	225
	21	128		24	169
	40	121		97	225
	71	169		51	169
	29	81			
	83	128			
	38	128			
	82	98			
	13	72			
	74	50			
	7	162			
	10	50			
	56	81			
	80	169			
	60	121			

Table 10

The (127,63,31) Difference Sets

There are six inequivalent difference sets with the parameters (127,63,31). Each is listed below in terms of nine numbers; for each j, the set contains all the numbers $2^i j \bmod 127$.

1) 1,9,81,47,21,31,25,49,15
   (the quadratic residues mod 127).
2) 1,3,9,27,116,87,7,124,15
   (the 6 dimensional projective geometry over GF(2)).
3) 1,3,27,87,7,63,25,75,5
   (the Hall set for primes of the form $4x^2 + 27$).
4) 1,3,9,27,81,87,7,31,25
5) 1,3,81,87,7,31,59,75,5
6) 1,3,81,29,7,21,63,31,25

Below is a listing of the sizes of the triple intersections of blocks of the designs they define:

Size of triple intersection \ Set	1	2	3	4	5	6
10	0	0	63	0	0	0
11	0	0	0	42	84	42
12	0	0	84	210	0	147
13	1197	0	840	630	1449	735
14	1890	0	1323	1806	0	1785
15	1323	7812	2772	2037	4284	2037
16	1890	0	1491	1848	0	1806
17	1323	0	819	924	1995	987
18	252	0	483	336	0	273
19	0	0	0	42	63	42
20	0	0	0	0	0	21
31	0	63	0	0	0	0

All other entries are 0.

REFERENCES

1. Baumert, L. D., Difference Sets, SIAM Meeting, Santa Barbara, 1967.

2. Boehmer, Ann M., Binary Pulse Compression Codes, IEEE Trans. on Information Theory, Vol. IT-13, 1967, pp. 156-167.

3. Brauer, A., On a New Class of Hadamard Determinants, Math. Zeit, 58(1953), pp. 219-225.

4. Bruck, R. H., Difference Sets in a Finite Group, Trans. Amer. Math. Soc., Vol. 78, pp. 464-481, 1955.

5. Frank, R. L., Polyphase Codes with Good Nonperiodic Correlation Properties, IEEE Trans. on Information Theory, Vol. IT-9, 1963, pp. 43-45.

6. Goethals, J. M. and Seidel, J. J., Orthogonal Matrices with Zero Diagonal, Can. J. Math., Vol. 19, 1967, pp. 1001-1010.

7. Golomb, S. W., Shift Register Sequences, Holden Day, 1967.

8. Golomb, S. W. and Scholtz, Generalized Barker Sequences, IEEE Trans. on Information Theory, IT-11, (1965), pp. 533-537.

9. Gordon, B., W. H. Mills and L. R. Welch, Some New Difference Sets, Canad. J. Math. 14(1962), pp. 614-625.

10. Hall, Marshall, Jr., Combinatorial Theory, Waltham, Mass. 1967.

11. ________, On Hadamard Matrices, SIAM Meeting, Santa Barbara, 1967.

12. ________, A Survey of Difference Sets, Proc. Amer. Soc., Vol. 7, pp. 975-986, 1956.

13. Kelly, J. B., A Characteristic Property of Quadratic Residues, Proc. Amer. Soc., Vol. 5, pp. 38-46, 1954.

14. Lunelli, Lorenzo, Tabelle di Sequenze (+1, -1) con Autocorrelazione Troncata non Maggiore di 2, Politecnico di Milano, 1965.

15. McFarland, R. and H. B. Mann, On Multipliers of Difference Sets, Canad. J. Math. 17(1965), pp. 541-542.

16. Mann, H. B., Addition Theorems, Interscience, 1967.

17. ________________, Balanced Incomplete Block Designs and Abelian Difference Sets, Illinois J. Math. 8(1964), pp. 252-261.

18. Menon, P. Kesava [1], Difference Sets in Abelian Groups, Proc. Amer. Math. Soc., Vol. 11, pp. 368-376, 1960.

19. ________________ [2], On Difference Sets Whose Parameters Satisfy a Certain Relation, Proc. Amer. Math. Soc., Vol. 13, pp. 739-745, 1962.

20. Poliak, Iu. V. and Moshetov, R. V., Scientific Transactions of the Radio-Engineering and Institute of the Acad. Sci. USSR 1; pp. 124-159.

21. Rankin, R. A., Difference Sets, Acta Arith. 9(1964), pp. 161-168.

22. Roth, R., Correction to a paper of R. H. Bruck, Trans. A.M.S., v119(1965), pp. 454-456.

23. Ryser, H. J., Combinatorial Mathematics, Carus Mathematical Monographs, 1963.

24. Singer, J., A Theorem in Finite Projective Geometry and Some Applications to Number Theory, Trans. Amer. Math. Soc. 43 (1938), pp. 377-385.

25. Stanton, R. G. and Sprott, D. A., A Family of Difference Sets, Can. J. of Math., V. 10, 1958, pp. 73-77.

26. Storer, J. and Turyn, R., On Binary Sequences, Proc. Amer. Math. Soc., Vol. 12, pp. 394-399, 1961.

27. Turyn, R. [1], Character Sums and Difference Sets, Pac. J. Math., 15(1965), pp. 319-346.

28. ________________ [2], The Correlation Function of a Sequence of Roots of 1, IEEE Trans. on Information Theory, Vol. IT-13, 1967, pp. 524-525.

29. ________________ [3], Ambiguity Functions of Complementary Sequences, IEEE Trans. on Information Theory, Vol. IT-9, 1963, pp. 46-47.

30. Yamamoto, K., Decomposition Fields of Difference Sets, Pac. J. Math., 13(1963), pp. 337-352.

31. Yates, R. D., Polyphase Sequences with Good Cyclic Correlation Properties, SIAM Meeting, Santa Barbara, 1967.

32. Zierler, Neal, Linear Recurring Sequences, Jour. SIAM, 7 (1959), pp. 31-48.

INDEX